BUILDING A BETTER BOAT

HOW THE CAPE ISLAND LONGLINER SAVED NOVA SCOTIA'S INSHORE FISHERY

DONALD J. FELTMATE

T0243964

NIMBUS
PUBLISHING
— NIMBUS.CA —

Nimbus Publishing Limited
3660 Strawberry Hill Street, Halifax, NS, B3K 5A9
(902) 455-4286 nimbus.ca

Printed and bound in Canada

NB1617

Editor: Marianne Ward
Editor for the press: Angela Mombourquette
Interior design: Jenn Embree
Cover design: George Kirkpatrick
Cover images: NS Archives

Library and Archives Canada Cataloguing in Publication

Title: Building a better boat : how the Cape Island longliner saved Nova Scotia's inshore fishery/Don Feltmate.
Names: Feltmate, Don, author.
Identifiers: Canadiana (print) 20220455619 | Canadiana (ebook) 20220455724 | ISBN 9781774711583 (softcover) | ISBN 9781774711590 (EPUB)
Subjects: LCSH: Fishing boats—Nova Scotia—History—20th century. | LCSH: Fisheries—Nova Scotia—History—20th century. | LCSH: Longlining (Fisheries)—Nova Scotia—History—20th century.
Classification: LCC SH344.8.B6 F45 2023 | DDC 623.82/8209716—dc23

Nimbus Publishing acknowledges the financial support for its publishing activities from the Government of Canada, the Canada Council for the Arts, and from the Province of Nova Scotia. We are pleased to work in partnership with the Province of Nova Scotia to develop and promote our creative industries for the benefit of all Nova Scotians.

This book is dedicated to the Nova Scotia fishers who

participated in the recovery of the shore fishery and

the development of the Cape Island–type wooden

longliner, a forgotten part of our Maritime heritage.

Let us honour their memory by remembering and

acknowledging their courage, perseverance, and

dedication in building a better boat.

CONTENTS

AUTHOR'S NOTE

I CONSIDER MYSELF EXTREMELY FORTUNATE TO HAVE BEEN BORN INTO a fishing family and to have grown up in that special period in our history when the inshore (commonly known as the "shore") fishery was reaching the apex of its recovery and the wooden Cape Island–type longliner was in its heyday. My father, both grandfathers, and paternal uncles were all fishers who, like many others, practised their craft in the Nova Scotia coastal and middle bank fishery. As young men, they lived through a period when the Nova Scotia shore fishery was going through some very hard economic times and was on the verge of collapse. It was the time that a large number of our coastal communities were living in severe poverty and in some cases on the brink of starvation. This difficult period of Nova Scotia's history has been largely ignored.

About the time I was born, mid-1949, the recovery was just underway. As I got older, I heard the stories of the "hard times," of the problems the fishers of the time were having with securing funds for new fishing craft, the very low fish prices, the incursion of large draggers on the coastal fishing grounds, and the lack of interest by both levels of government, provincial and federal. Unfortunately, I was too young to put the pieces of this incredible story together.

As a young boy growing up in the Cape Breton village of Port Morien, I always had a keen interest in the activities of the local wharf. I can still remember the local and visiting swordfishing boats—the forerunners of the Cape Island–type longliner—and the excitement of watching each wander back to port in the early evening. I can remember seeing my first longliner. I always marvelled at their size in comparison to the smaller inshore boats and revered those who owned and sailed them; they were like demigods to me. For many of my friends and myself, these were the "super boats," the real thing, and as boys we dreamed of someday owning one. Fishing from it remained in our every thought. Visits to ports like Louisbourg, Glace Bay, Port Bickerton, and others prompted me to start remembering the names of the vessels, many of whose owners/skippers were dear friends of my family members. They were all part of the fraternity called the commercial shore fishery. Names like *Leaside, Bonny Lou, Danny B, Robin Lynn, White Whale, Debbie and Maxine, Arthur Ross,* and of course, my uncle's *Jean Elizabeth* were imprinted on my mind, and I would always go looking for them each and every time I visited their home port.

As I grew older and went fishing with my family members, I soon became familiar with the hard work and hardships that were faced by the fishers of the time. My late father and both my grandfathers would tell the stories of what it was like "back then" and talk about the struggles they had endured. I loved hearing those stories, but again, I could not really connect the dots between the troubles with the shore fishery, the need for better fishing platforms, and the significance that a new class of fishing vessel called the longliner had and continued to have on the shore fishery. Life was not easy for those who chose this profession. I remember the grief of a schoolmate in the early '60s when she lost her father in the tragic sinking of the longliner the FV *Elizabeth and Leonard* out of Glace Bay. I experienced first-hand the worry and loneliness that my aunt felt during those periods when my uncle was at sea, especially during the long and hazardous trips to the fishing grounds in the winter months with his longliner the FV *Jean Elizabeth*. These vessels not only added a new dimension to the shore fishery, they added a different mindset as well. No longer was the shore fishery defined as an industry conducted close to shore; it was expanding to the grounds that were out of reach of the smaller coastal vessels of the past. The fishers now had a tool that allowed them to fish the middle banks that extended twenty-five miles from the coast and beyond, stay at sea for longer periods, and fish later into fall and winter. This change resulted in risks that were not normally associated with day fishing.

By some strange twist of fate, I never did follow fishing as a career, but I never once forgot it—especially the longliners. Over the years, my love for this vessel never wavered. With the passage of time, I learned I was not the only one who shared this affection. I still remember the host of aging coastal fishers who shared my fondness and love for this fine fishing craft and whom I would see on my treks to the wharves.

FV *Jean Elizabeth* at Whitehead, NS, covered with ice after surviving a major storm that claimed the lives of seventeen fishers from Lockeport, NS, March 18, 1961. (AUTHOR'S COLLECTION)

Each time I returned to my childhood home, I made a pilgrimage to the wharves in Glace Bay, Louisbourg, and other ports looking for my old familiar "friends." Sad to say, they were getting harder to find, and like me, those who remained were getting older and "showing grey around the edges." On a visit to Glace Bay in 1994, I suddenly realized that the longliners—the vessels I had come to know and revere—were finally gone. I also realized, to my dismay, that the captains who sailed and fished these magnificent craft were well up in their years, and sadly a good number were no longer with us. Life had moved on and there was no returning to the past.

It was about that time I started inquiring to see if anyone had had the foresight to honour these vessels by compiling a list of those that fished our local waters and even more important, recorded the story of how they came to be and the role they played in the economic prosperity of Nova Scotia. To my surprise there was nothing. There are stories and tales and some conjecture about specific events, but there was nothing in print that recorded and offered a précis of the complex history of the shore fishery and the development of the Nova Scotia Cape Island-type longliner.

In 2009, I made a decision to research and record the history of the Cape Island–type longliner. With the sponsorship of the Fisheries Museum of the Atlantic in Lunenburg, NS, I was able to find and catalogue the history of 205 of these fine craft. This resulting document, "The Cape Island Type Longliner: A Nova Scotia Legend, Volume Two, A Pictorial History 1949–1985," contains the history of each of the 205 fishing craft that are classified as the wooden Cape Island–type longliner as defined by the Department of Fisheries in 1947. There is a possibility that there were others, but with closure of public access to the official registers in 2013, it was all but impossible to verify any additions. (The document, which was never formally published, resides at the museum. There was originally a notion to have a companion document—hence, "Volume Two"—but it never materialized.)

As I moved forward with this quest, I discovered that the story of the longliner is not just about another type of fishing vessel but is also about the inshore fishery itself, and that the longliner was interwoven with the fishery's very existence. The stories I had heard as a young boy started to match those contained in the documents I was researching and drove me to dig deeper. The development of the Cape Island–type longliner was perhaps one of the most significant achievements in the struggle of the inshore fishery during the first half of the twentieth century and played a major role in saving the coastal fishery from almost certain collapse. It brought about a period of prosperity to a host of villages and small towns along the Nova Scotia coast, and it is perhaps safe to say that a number of small coastal Nova Scotia villages would not exist today had it not been for the Cape Island–type longliner.

The Cape Island–type fishing vessel had its beginnings in 1905 in Clark's Harbour, Cape Sable Island, NS. In that year, a well-known local carpenter and boat builder by the name of Ephraim Atkinson designed and built a small lobster boat that broke away from the traditional schooner hull design. The idea of placing an internal combustion engine in a fishing vessel was in its early infancy, and the placing of a small, single-cylinder engine in the conventional sailing craft of the time was not without its difficulties. The engine, as rudimentary as it was, made the small sailing craft difficult to handle when tending lobster traps, etc. Mr. Atkinson realized that for small fishing craft to accommodate an engine, there had to be changes so the vessel would handle better. His design broke away from the traditional rounded stern and narrow bow found in the schooner design of the time and incorporated a more rounded bow, an increased width, and a flat, wider stern. The vessel had a greater sheer than the conventional sailing craft and proved to ride the coastal swells and handle much better than any sailing craft of the time. Some of the older fishers say Atkinson designed the hull after watching a seagull ride the waves. His small boat proved to be an excellent craft for the coastal lobster and groundfish fisheries, and it was not long before other craft stemming from this design started appearing in the local fishery. Over time other builders started to improve on this design, making it the most identifiable fishing craft in the Atlantic fishery. It was from this humble design that the Cape Island–type longliner would evolve some forty years later. There is no question that the development of this type of fishing craft was as significant to Nova Scotia as the famous Grand Banks schooners that, not unlike the longliner, evolved from smaller coastal sail-powered fishing craft.

Nova Scotia is the only Maritime province that through a process of need and evolution designed a specific type of fishing vessel, developed the architectural drawings to conform to international standards and conventions, had the design designated as a separate class of fishing vessel, and built a vessel that was uniquely Nova Scotian in all respects. Its introduction into the fishery came at a critical time and provided an affordable fishing craft to the individual fisher that could safely address the needs of the shore fishery.

Sadly, this very important aspect of Nova Scotia's history has gone largely unnoticed. This book offers a summary of major events, based on historical documentation and interviews with those who lived through and were part of the difficult times in the inshore fishery. It is the story of a struggle by those who had dreamed of a better fishing craft. It is the story of those who built and fished this class of vessel and more importantly demonstrated how this vessel came to be. It is a story that needs to be told, if for no other reason than to honour those brave men and women that persevered through unimaginable adversity, who believed in Nova Scotia and that there was a future in the fishery. Against all odds, they developed a "tool" for their profession. A tool that not only became recognized internationally, but also touched the lives of every Nova Scotian in one way or another.

INTRODUCTION

*I loves the Cape Islander. I got to say they
got to be the best lookin' type of craft ever
built, not countin' their seaworthiness.*

– LLOYD NAUGLE, FISHER

THE HARVESTING OF GROUNDFISH AND OTHER SPECIES HAS ALWAYS
played a prominent role in the economic viability of Nova Scotia,
dating back a century before Confederation. It was the rich fishing
grounds near and adjacent to the Nova Scotia coastline that led to
its settlement. Beginning in the 1700s, fishing was a way of life for
a large portion of the population and was only matched as a major
source of employment and prosperity by the lumber and shipbuilding
industries of the early 1800s. A survey conducted by the Department
of Fisheries and Oceans in the mid-1960s found that 75 percent of
all the communities in the Atlantic Region had fishing as a base for
employment, and 20 percent were wholly dependent on that industry.

Many stories have been written about the glory days of the Grand
Banks fishery when the handsome schooners like the famous *Bluenose*
and its fellow vessels sailed from major ports such as Lunenburg,
Canso, Digby, and others with their iconic dark, sleek hulls and bil-
lowing sails capturing the imagination of all who saw them. Even
today the historic images of the schooners of this era are identifiable
throughout the world as being part of Nova Scotia's seafaring heritage.

It is a little-known fact that during this same period there was
another fishery—the shore fishery—that had no glory and no large
schooners. What is unknown to most and almost forgotten by

others is that in the early part of the twentieth century, thousands of hard-working fishers, the majority of those engaged in the Nova Scotia fishery, were living in small fishing communities in extreme poverty, facing starvation, and barely surviving. Even those who were fortunate enough to find employment in the offshore or banks fishery were themselves poorly paid, and their living conditions, not to mention the dangers of their profession, were less than ideal and not much better than the shore fishers.

By the early 1900s, things began to change and for a host of small fishing communities this change was not good. After the end of the First World War, there was a major economic recession that had a big impact on the Atlantic fishery, with the hardest hit being the inshore fishery. The majority of these coastal communities were stricken with despair and a hopelessness that began to break down their social and moral fabric, and what is most appalling is that neither the provincial government nor the fledgling federal government seemed to care! Their story has been forgotten, yet it remains a very important part of Nova Scotia's seafaring heritage.

From the early 1900s until the mid-1930s, the shore fishers were at the mercy of the larger fish companies and other processors that were the major purchasers of their catch. Some of these companies either owned or had shares in the now-famous fishing schooners. During this period, Canada claimed territorial jurisdiction over the waters within three nautical miles of its coastline. The three-mile limit did not extend any protection to the fishing grounds used by the shore fishers and allowed large steam trawlers or "draggers" to operate on the same fishing grounds. By 1920 the incursion of the steam trawler was having a disastrous effect on the traditional near-to-shore fish stocks. The ever-dwindling catches in the coastal waters caused a sharp decline in the income of the shore fishers and forced them to go farther out to sea. Unfortunately, most did not have the resources to purchase the larger vessels or the equipment needed for this task.

In 1927 in the town of Canso, a number of Nova Scotia fishers peacefully and openly resisted the status quo and challenged both levels of government to recognize their dire social situation and provide some support to the shore fishery. Their actions set off a period of political turmoil that would last for the next twenty-five years. The problems that brought this segment of the Nova Scotia fishery to the verge of total collapse in the late 1920s are also integrally tied to its recovery, the introduction of a new fishing method known as longlining, and the development of a medium-sized Nova Scotia fishing craft known as the Cape Island–type longliner. The first wooden, government-approved Cape Island–type longliner appeared in a Nova Scotia shipyard in 1950, more than twenty years after the fishers' revolt of 1927. This was the amount of time that passed before shore fishers started to see any sizable movement toward their financial recovery. This sturdy little fishing boat was not the brainchild of any single naval architect but was the direct result of a series of economic, social, and political events that began two decades before. It would sail on to become a key element in saving and revitalizing the shore fishery.

Those who were not brought up on the sea are no doubt wondering, "What is longlining and what is a longliner?" The definition of the term "longlining" cannot be found in any nautical references such as the *Oxford Companion to Ships and the Sea*. It is a term unique to the fishing industry, used to describe a method of fishing groundfish like cod, haddock, pollock, and similar species using a line buoyed at each end, moored on the fishing ground, and with smaller lines, each with a baited hook, attached at short intervals along its length. This method of fishing was used in Europe a good century before it was introduced to the Canadian fishery, and it may surprise some to learn that it was used with considerable success in the West Coast fishery long before it was accepted on the East Coast.

In the days of the well-known Grand Banks schooners, trawl lines of this configuration were employed from dories that were carried on board the larger schooners for that purpose. Usually each dory carried a two-person crew and set out from the main ship with three tubs of trawl lines. The lines would be set out and hauled back by hand, and the catch would be returned to the main ship where it would be cleaned, salted, and placed in the hold. This was dangerous and back-breaking work that was conducted during fishing trips to the banks that lasted up to two months. This method for catching groundfish remained the domain of the offshore fishing fleet that is historically known as the "salt bank fishery."

Unlike the "salt bank" or offshore fishery of the time, the "shore" or inshore fishery was conducted on the coastal fishing grounds adjacent to the Nova Scotia coast. The boats used in this segment of the fishery were quite small, mostly powered by sail or oars with extremely limited range and seaworthiness. The most common method for harvesting groundfish was the handline, which uses a single hook. In some fishing communities, trawl lines of a sort were used, but the amount of gear was limited by the size of the vessel and the ability of the fisher to purchase such equipment and bait.

With the introduction of larger internal combustion engines in the early 1920s and the mechanical hauler in the early 1930s, the days of the schooner were numbered, and the use of dories diminished as the trawl lines could now be set and hauled from a single vessel. Thus, each set of trawl lines became significantly longer and became known as "longlines." It stood to reason, therefore, that the vessels employed in this type of groundfishing became known as "longliners," a term that remains to this day.

To most Nova Scotians who are familiar with the fishery, the term "longliner" applies to any type of vessel that engages in groundfishing with a number of tubs of trawl lines. Many would be surprised to learn that from 1942 to 1950, after considerable political turmoil and controversy, the federal and provincial governments established a clear definition for the specific type of vessel called a "longliner," and in the case of the wooden Cape Island–type longliner, they further defined it as a separate class of fishing vessel. Another surprising fact is that the classification of this type of fishing was not solely based on how the vessel was to be employed, but also on how the vessel was constructed and whether it met the qualification criteria for financial aid from the federal government in the form of subsidies. The traditional Grand Banks schooner gave way to the motor dragger, and longlining became one of the prime fishing methods of the small fishing communities along the Nova Scotia coast. The problem was that most shore fishers did not have fishing craft capable of engaging in this technology.

The Cape Island–type longliner was a major product of a very tumultuous period in Nova Scotia's seafaring history and brought with it a measure of economic recovery and prosperity to a number of small and medium-sized communities along the coast. The economic impact that these vessels had on the life of every Nova Scotian between 1950 and 1985 has hitherto gone virtually unnoticed. The story of the shore fishery and the eventual introduction of the longliner is one of an incredible journey in the economic and social development of Nova Scotia. It is one of heartache, political indifference, and perseverance. The design and introduction of the longliner to the inshore fishery was matter of survival for a large population of our shore fishers, their families, and the communities that were home to generations. This book is an attempt to document the journey of the shore fishers from the dark days of the late 1920s, through the political upheaval of the 1930s and '40s, until the last known wooden "government approved" Cape Island–type longliner was built in 1984. Based on archival documentation and interviews with those who, during this period, built these magnificent craft and a number of the remaining shore fishers that sailed them, what follows is the forgotten story of how a group of ordinary Nova Scotians, with a dream of a better future for their families and communities, overcame insurmountable odds to improve their social and economic situation and in the process develop, build, and sail an iconic Nova Scotia craft. Let the journey begin.

CHAPTER 1

IN THE BEGINNING

*You see, God Almighty knows everything
and when He made the fisherman, He knew
that there was a certain class of people out
to skin him. So, He took precautions and
ordered it so that when one skin was pulled
off him, another grew on. It's a good thing
He did or 90 percent of the fishermen today
would be walking around skinless!*

– CAPTAIN WILLIAM (BILLY TOM) FELTMATE

THE LAND MASS KNOWN AS CANADA IS LOCATED ADJACENT TO SOME of the richest fishing grounds in the world. Europeans, including the English, French, Spanish, Portuguese, and Basque, began fishing off the coasts of Nova Scotia and Newfoundland in the sixteenth century. Commercial fishing along the coast of British Columbia began with the settlement of that colony in the mid-1800s. Use of the valuable fishing grounds on both the Atlantic and the Pacific coasts continued to grow, and by the beginning of the 1900s, vessels from almost every major maritime nation in the world could be found fishing off the coasts of Canada. Most of the maritime nations did and still recognize the importance of the fishing grounds off Canada's coastline and the impact this fishery had and continues to have on their economy. It is not surprising that these nations invested heavily in their fishing industry. By contrast, in the years following Confederation, the fledgling Government of Canada did absolutely nothing to support the nation's fishing industry.

This small fishing village was typical of those that dotted Nova Scotia's coastline. (NS ARCHIVES)

Of the two Canadian coastal fisheries, it was off the East Coast that the richest fishing grounds in the world could be found. In 1945 the waters off the East Coast of Canada available for fishing encompassed over 8,050 kilometres (5,000 miles) of coastline stretching from the United States border near Grand Manan Island, New Brunswick, to Labrador.[1] The Bay of Fundy, located between Nova Scotia and New Brunswick, alone has a geographical area of 20,720 square kilometres (8,000 square miles) and in itself is a rich fishing ground. The Gulf of St. Lawrence includes an area ten times larger, and other ocean waters, such as the coast of Labrador and the eastern Arctic, make up an additional 516,000 square kilometres (200,000 square miles) or four-fifths of the total fishing grounds of the North Atlantic.[2] It is therefore ironic that at the beginning of the twentieth century and despite having a more favourable geographic position in relation to the bounty of these rich fishing grounds, Canada held tenth place among the world's larger fishing nations. What is even more ironic is that just prior to the First World War, the relatively small British colony of Newfoundland had a larger and more productive fishing industry than did the collective fisheries of the three Canadian Maritime provinces and British Columbia.

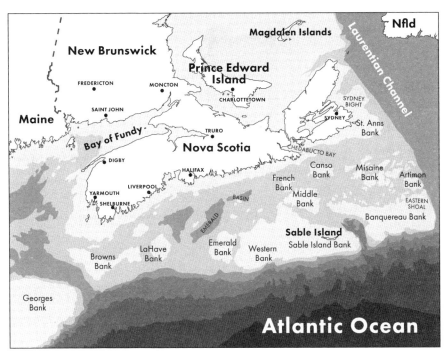

Major fishing banks and basins on the Scotian Shelf.

Within the Canadian fishing industry there are two very distinct fisheries, each individually driven by geography, organization, profitability, species of interest, and fishing methods. These are the Atlantic fishery and the Pacific fishery, often called the West Coast fishery. The West Coast fishery of the early 1900s was more highly organized with more modern processing facilities and a unified marketing strategy. As an example, as early as 1870, the processing plants along the British Columbia coast had fully industrialized the salmon fishery by building a number of very efficient canneries and worked collectively to establish sound and profitable markets for their product. The size of the vessels varied depending upon whether they were employed in the coastal or offshore fishery, with the larger number being in the coastal fishery. The fleet was well maintained and kept pace with any technological advancement that would benefit their industry. What is significant is that, unlike the East Coast fishery, by 1920 the majority of the West Coast fleet was motorized, while as late as 1936, Nova Scotia was still building sailing schooners.

When the West Coast fishing industry was first established, the fishing centres that included small communities and processors located their facilities close to the fishing grounds.[3] The industry itself was concentrated in specific centres and co-located with the processing plants and supporting infrastructure. This situation did not occur as a result of any direct federal intervention or planning but rather the physical geography of the British Columbia coast and more specifically the role Vancouver Island would play in the fishery. The fishers themselves were very organized and had their own union and collective bargaining structure that allowed them to act with a spirit of solidarity when establishing markets and catch prices. The relationship between the fishers and the processors worked reasonably well, and both were able to profit and have some capital available for the purchase or upgrade of existing facilities and fishing vessels to meet the changing needs of the industry.

The structure of the fishing industry on the East Coast from 1900 until the mid-1930s was directly the opposite. The Atlantic fishery was divided among the four separate provinces that bordered on the Atlantic Ocean or the Gulf of St. Lawrence, the French colony of St. Pierre and Miquelon, and the British colony of Newfoundland. The problems associated with the geographical and ethnic diversity of the region made any attempt to establish a viable fisheries association very difficult. This situation was further exacerbated by the fact that at Confederation, Quebec was granted complete jurisdiction over its own fishery while the federal government was responsible for the remaining provinces of Nova Scotia, New Brunswick, and Prince Edward Island.[4] Therefore, the ability of the Atlantic fishery to speak with a single voice was nonexistent, and the problems that ran rife within the industry were essentially invisible to the federal government.

The Nova Scotia fishery comprised two components: the offshore and inshore. In the beginning, both fisheries were closely aligned, with the offshore fishery concentrating its operation in the middle fishing banks region just beyond the coastal waters of Nova Scotia. By the end of the First World War, the inshore and offshore fisheries in Nova Scotia started to emerge as two very separate and distinct fisheries. The offshore fishery was not defined by the distances the vessel had to travel to the fishing grounds but was predicated on the fact the vessels remained on the grounds for several days to several weeks before returning to port. With the passage of time, the offshore fishery became increasingly more structured and centred its activities in the major ports such as Lunenburg, Digby, Lockeport, Shelburne, and Canso. Each of the ports that supported the offshore fleet had one (or more) major fish processing/handling facility and a fairly good market for fish landings. The vessels engaged in this fishery were normally medium to large schooner-type craft with the capability to travel to the major East Coast fishing grounds and remain there for extended periods. The offshore fishery was therefore relatively stable, and as a result, each of the ports had over time developed the infrastructure to support the vessels and the landings. Ports like Lunenburg, for example, had access to a reasonable network of roads and railway lines that could be used to transfer processed product to market. Integrated with the port were shipyards, chandlers, and outfitters that facilitated the building, maintenance, and repair of the offshore fleet. Perhaps more important was that the major fish processing facilities had the capital to adjust their processing capabilities to suit a changing marketplace.

In Nova Scotia, the inshore fishery was conducted relatively close to the coast out of a multitude of small ports and harbours along the province's coastline. The fishing craft engaged in this fishery were quite small, of open construction, powered either by sail or oar, and returned to port with its catch every day. Most fishers operated within six nautical miles of the coast. By the early 1920s the gasoline engine began to make its appearance in fishing vessels, and the designs of the inshore boats were seeing changes; however, the cost of purchasing such craft was out of the financial reach of most. A number of these small ports, like Whitehead, Larry's River, Woods Harbour, Grand Étang, and others, could be classed at the time as being semi-isolated, as the overland transportation links to these communities were rudimentary at best. Most of the small settlements did not have processing plants or the necessary infrastructure to support any expansion or consolidation of markets. Goods and services came via the "freight boat" that serviced the small communities on a regular basis from major centres like Halifax and Sydney.

At the beginning of the twentieth century, lobster was still the major species of interest and made up the bulk of the "cash crop" for the inshore fisher, albeit the prices being paid by the local buyers were very, very low. Depending upon the area and the season, swordfish, cod, haddock, mackerel, and herring were also major contributors to the inshore fishery. Unlike the offshore fishery where the market at each of the major ports for landed fish was fairly stable year-round, the inshore fishery was primarily a seasonal affair, and the species varied not only with the season, but also by geographical area within the province. In some areas such as the South Shore of Nova Scotia and the eastern shore of Cape Breton, lobster was considered the backbone of the local fishery, while in other areas cod, mackerel, or herring was the major species of interest.

At this time in Nova Scotia's history, the groundfish caught was salted and prepared for market by the individual fisher. In some communities where there was a resident buyer, some fishers chose to sell their catch direct to the buyer, which further lowered the price. In the small communities away from the major centres, the fisher was at the mercy of a local fish buyer or a travelling broker commonly called a "smack," and it was not uncommon to have a large differential in the prices being paid for their catch versus those being enjoyed by fishers in the larger offshore centres.

The quality and quantity being landed by the offshore fleet also impacted the markets for the inshore. Sometimes the offshore fleet could singularly meet the market demand, thus driving the price paid to the inshore fisher even lower. Unfortunately, the subsistence of the inshore fisher became a system of barter where fishers bought most of their supplies, gear, and living staples from the buyer on credit and sold fish to that buyer. Having an existing debt, the fisher was at the mercy of the buyer and had to accept the price paid regardless of the quality of his fish. Quite often the fisher saw very little cash in hand and became subservient to the buyer. This system of barter became a normal way of life for the inshore fisher. In 1936, the Rev. Dr. Moses Coady, in a brief to the Honourable J. E. Michaud, minister of fisheries, eloquently summed up this system.

It is primarily a system of barter from which bargaining power has been almost entirely eliminated, in which unsound credit has probably played a more vicious role than in any other primary industry, and by which the fisher benefits little more, if any, from the good years than bad ones. If at times favourable markets prevail, somewhat higher prices may be paid for fish, but so invariably must higher prices be paid for equipment, food, and clothing. When a seasonal operation results in a small surplus, this must usually be applied to a debt. What happens more frequently, however, is that a new debt is added to the old one. This goes on until the merchant must need to protect his interest by taking as security homes, vessels and gear, if these are not already encumbranced, and so the fisher, through a combination of circumstances becomes only another instrument for the production of wealth for others instead of remaining a primary producer in a position to command a fair share of the value of the articles produced by his labour and investment.[5]

Between 1867 and 1918, the fishers scattered along the coast of Nova Scotia were able to subsist on their catches and in some areas stretching from Tancook Island to Yarmouth were for the most part doing quite well. However, in 1910, a new type of fishing vessel with a new technology started to appear in increasing numbers off the Nova Scotia coast, and the Nova Scotia fishing industry was changed forever. The steam trawler was a large steel-hulled vessel with an average length of 100 feet (30 metres) that dragged a large net known as an "otter trawl" along the floor of the fishing grounds taking everything in its path and catching more in a single haul than a schooner could do after weeks of fishing with the traditional hook-and-line methods. Most of these vessels came from the United States, but vessels from European countries started to fish on the near shore grounds that had been the traditional domain and backbone of the Nova Scotia shore fishery. Traditional stocks started to disappear, and the fishers were required to operate farther from shore thus requiring larger craft, the purchase of which was beyond the financial means of most.

By the mid-1920s and despite the overwhelming number of people involved, the inshore fishery was rapidly becoming very ineffective. The number of seasonal fishers engaged in the lobster industry was far greater than those engaged in the fishing industry on a full-time basis, a situation that remains to this day. This was particularly evident in the small ports around the industrial centres of the province. Of the smaller number engaged in the fishery on a full-time basis, 75 percent barely subsisted on their earnings from the sea. During this period, it was not uncommon for fishers in the ports some distance from the major centres of Glace Bay, Sydney, and Halifax to seek other work to supplement their income just to survive. Young people left the sea to seek employment in the mines or the expanding manufacturing sector of central Canada and the United States. It is therefore not surprising that between 1890 and 1927 the number of full-time fishers in Nova Scotia declined from 28,224 to 16,127. Most of this decline occurred in the inshore fishery.

Most Nova Scotia coastal fishers, although masters of their craft, had very little formal education, and a large number were illiterate. Among the inshore fishers in Nova Scotia, there was a fervent resolve to retain their individual independence and seek prices singularly rather than as a unified body with a single voice. Between 1920 and 1933 this independent spirit and lack of formal education placed the inshore fisher at a distinct disadvantage when trying to compete in a changing market and almost resulted in the collapse of this segment of the industry.

Prior to the First World War, Nova Scotia held a very lucrative market for salt fish with the West Indies, the United States, and some European nations. After the war, Norway and Iceland placed heavy emphasis on their fishing industry and started a program of major expansion and aggressive marketing. The fishing industry in these countries was being highly subsidized by their respective governments, and within a very short time they, along with Newfoundland, began to challenge and become extremely competitive in the Canadian "dry fish" market. Between 1920 and 1923, the Norwegian government paid $600,000 to secure the Havana salt fish market. This competitiveness and ability to provide a better quality product at a more reasonable price resulted in Nova Scotia losing a significant portion of this traditional market. The downturn in the salt fish industry and the inability of Nova Scotia to adjust to the changing market started to take a devastating toll on the fishing industry.

In 1920, 80 percent of the groundfish landed in Nova Scotia ports was sold to the salt fish market. By 1939, this amount had dropped to only 54 percent while the fresh and frozen market had risen from 9 percent to 34 percent. The offshore fishery, although hit hard by this radical change in market conditions, was able to introduce the necessary changes to fishing methods, fishing equipment, and processing plants to take advantage of what markets were available and enter new ones when they arose. These changes were possible only because the offshore fishery was tied to major fishing ports that had large processing facilities with the financial means to adjust to the changing marketplace. This was not the case for the inshore fishery, and this loss of the salt fish market was devastating. The remoteness of some of the small fishing ports, the lack of any facility to process fresh fish, and the lack of refrigeration and viable means of transporting their catch to larger facilities placed the inshore groundfishery close to the point of collapse. In a number of these small ports, lobster, herring, and swordfish became the only fisheries worth following.

During this same period, other changes were starting to take place in the food industry as a whole, and these changes were to have a significant impact on the fishing industry in Nova Scotia and in particular, the inshore fishery. The food industry in Canada and the United States started a program to develop and implement standards regarding the grading and quality of food products, and a new pricing system was slowly being established in the marketplace. The price paid for the product was dependent upon the grade and quality. The shore fisher was the producer and manufacturer of his or her product and once again was at the mercy of the buyer who determined the quality and normally paid the lowest price regardless.

Coinciding with the implementation of grading and quality standards was an ever-increasing demand for fresh foods, including fish products. This caused a major shift in the preparation of product for the market. Fresh fish needs more care in the handling and preparation for market, thus the old practice of shipping fish loose and in bulk was giving way to the requirement for individual packaging. The larger fish processors were now packaging, canning, and smoking fish products in an effort to attract new buyers. This shift in market and consumer needs caused the international fish market to become increasingly more competitive, and most major processors were taking measures to cut costs, improve production, and increase the quality of their product, initiatives that were well beyond the reach of the individual shore fisher.

Technology was also having its impact on the fishery as a whole. The introduction of refrigeration provided the processors a means to freeze produce, including fish products, as well freeze fresh bait. The consumer now had access to a higher quality product, and as a result, salt-cured products such as fish began to decline in popularity.

The era of the fishing schooner was slowly giving way to the steam trawler. In the early 1920s, the number of steam trawlers operating on the inshore fishing grounds continued to increase and do so without any oversight or restrictions. It was not uncommon for these vessels to visit Canadian ports such as Canso, Louisbourg, and Lunenburg. It did not take long for the major fish dealers to realize the dragger could catch more fish in a single day than a schooner could in a month on the banks, could be staffed by smaller crews, and could return to port quicker with no requirement to salt the fish on board. As a result, the large companies could now take advantage of the fresh fish market as well. This "new technology" proved to be very efficient, and the larger fish companies started to invest in the purchase or construction of trawlers.

By 1926 there were eleven Canadian- and a large number of American-owned steam trawlers working from one or more of the major Nova Scotia offshore fishing ports. Given that Canada's territorial water extended to only three miles offshore, it is not surprising that trawlers continued to operate on the small coastal banks very close to and sometimes on the inshore fishing grounds and quite often could be seen from the small coastal fishing communities. The inshore fishers watched helplessly as these vessels started dragging over their traditional grounds and catching more fish in one day than the average inshore fisher could in the course of a year. The heavy nets used by the steam trawlers were degrading the ocean floor, and valuable habitat for other species such as lobster was being destroyed. More importantly some of the draggers were towing their gear through the nets and traps of the inshore fishers. The buildup of the dragger fleet just off the coast of Nova Scotia was disastrous. The Nova Scotia inshore fishers simply were unable to compete, and the resentment by the inshore fishers for the offshore fleets continued to build.

Within the federal government structure there was no separate department of fisheries. At Confederation, fisheries fell under the responsibility of the federal minister of marine and fisheries. With the formation of the Royal Canadian Navy in 1910, this ministry was expanded to include the naval service. The structure remained so until 1920 when the naval service broke away to a separate ministry and the title reverted back to the Department of Marine and Fisheries. It was not until 1930 and as a direct result of the recommendations put forth by the Royal Commission Investigating the Fisheries of the Maritime Provinces and Magdalen Islands, known as the MacLean Commission, that the federal Department of Fisheries was finally established in Ottawa.

A similar situation existed within the provincial government structure of Nova Scotia. At the time of Confederation, fisheries were the responsibility of the minister of commerce and natural resources and remained that way until the late 1940s. The failure of both the federal and provincial governments to recognize the importance of the fishery to the Maritime economy and the reluctance to establish a separate department of fisheries opened both levels of government to resentment by the fishers and public ridicule by the press. From 1920 until 1929, *The Canadian Fisherman*, a monthly news publication founded in 1914 and dedicated to the fisheries, became the voice of the fishers for the establishment of a separate federal fisheries department. In the April 1921 issue of *The Canadian Fisherman*, the following advertisement appeared: "Wanted by the Canadian Fishing Industry, a Department of Fisheries separate and distinct from Marine, Naval and other affiliations. Also, a Deputy Minister in charge who will have direct access to the Minister."

By the mid-1920s the frail framework and scope of responsibilities of the federal Department of Marine, Fisheries, and Naval Service and the failure of the federal government to understand the importance of the East Coast fishery to Canada, almost caused the collapse of the East Coast inshore fishery. It was indeed unfortunate that during this period the federal Department of Fisheries did little to assist the Nova Scotia fishery. The focus of the federal government just prior to the Great Depression was the buildup of the manufacturing centres of western Quebec and southern Ontario. There was an ever-increasing number of initiatives to seek international markets for and to promote Canadian manufactured goods in the areas of textiles, machinery, and the automotive industry. Incentives were being given to entrepreneurs who wished to establish manufacturing facilities in central Canada, yet no consideration was ever given to the plight of the East Coast coastal fishery.

By the end of 1926, the bitterness and frustration of the inshore fishers along the Atlantic coast of Nova Scotia was reaching a boiling point. Stricken with a continual decline in traditional markets, poor fish prices, and lack of capital to improve fishing gear and methods, the fishers were being forced to the point of despair. In that same year, the Nova Scotia fishery was dealt another blow. The fiscal and trade policies of the federal government forced the introduction of tariffs on most manufactured goods made for export. Unfortunately for the fishing industry, processed fish products in any form fell under the definition of "manufactured goods" and thus were subject to the same regulation as any other item available for export. Some of these tariffs remained on goods crossing provincial borders, with certain types of seafood being among the items subject to taxes. To make a bad situation even worse, some of the tariffs varied by area and region within the province, causing a further isolation of fishers to a common market. The Nova Scotia shore fishery had reached its lowest point in the history of this young nation and was on the very verge of total annihilation.

CHAPTER 2

THE PRIESTS

*Tompkins found debt-ridden fishermen living
in weather-beaten shacks. They sold their
catches wherever they could for a fraction
of their value. In bad times they mortgaged
their homes and boats to feudal fish
merchants and storekeepers. Their children
were underfed, poorly clothed. They were
dejected and defeated.*

– DAVID MACDONALD, *MACLEAN'S* MAGAZINE, JUNE 1, 1953

BEGINNING IN THE MID-1920S, THE POVERTY AND DESTITUTION THAT would have matched that of any developing nation had become a way of life for a large number of the small fishing villages along the Nova Scotia coast. Similar conditions existed in the other Atlantic provinces. Most of the families were finding it hard just to survive, and any hope they held for improving their lot was slowly beginning to fade. The independent inshore fishers were totally fed up with the lack of action by both the federal and provincial governments to provide some direction and support with respect to their treatment by larger fish dealers, the incursion of foreign steam trawlers on their traditional coastal fishing grounds, and the lack of any financial support for the purchase of better vessels and equipment. They were not asking for governmental handouts but a methodology that would allow them to move forward and allow them to make an honest living from the sea such that they could proudly care for their families and contribute as citizens to this new country called Canada. Unfortunately, this did not happen and the majority of the shore fishers fervently believed that they had been betrayed by Nova Scotia's entry into Confederation and were now alone in their struggle to exist. With Confederation, the federal government now had control over the Nova Scotia fisheries, and this did not sit well with the inshore fishers. All along the shores of Cape Breton, the Eastern Shore, and portions of the South Shore, the residents of the small fishing communities began to question why Nova Scotia had ever entered Confederation, and the seeds of a separation movement of sorts were slowly taking root. Similar conversations were surfacing in the Acadian communities along the Bay of Fundy. This air of despair and loneliness was further compounded by the semi-isolation of the communities brought on by the lack of direct landline communication and an acceptable system of roads. The citizens of these communities believed they were somehow being singled out and forgotten by those who had the power to help. Despite this very dire situation in Nova Scotia, they were not alone in their struggle. Similar conditions existed in New Brunswick, Prince Edward Island, and the Magdalen Islands.

Unbeknownst to maritime inshore fishers, the seeds of recovery had already been sown and started to take root in the small Nova Scotia fishing town of Canso in the form of their parish priest, a fiery Cape Bretoner, Rev. Jimmy (J. J.) Tompkins, and a social visionary— the Reverend Dr. Moses Coady. To understand how this recovery was initiated and the impact this would have on the development of the longliner class of fishing craft twenty years hence, it is important to examine what was taking place in the whole geographic area of Eastern Nova Scotia and Cape Breton at the time and have some understanding of the growing social movement that was to become known both in Canada and abroad as the Antigonish Movement. In the case of the shore fishery, some believe it should have been called the Canso Movement.

As the fledgling nation of Canada entered the twentieth century, the fishery was not the only industry that was experiencing difficult times. Like the shore fishery, the agricultural sector, especially in the Cape Breton Highlands, was struggling to survive. The regional out-migration of young people that started at the turn of the century continued unabated into the 1920s, as young, able-bodied farmers left

Rev. Dr. Moses Coady and Rev. James (J. J.) Tompkins were pioneers of the Antigonish Movement, which helped sow the seeds of recovery in the inshore fishery.

Above: Rev. Dr. Moses Coady (KIDEN-KAZANJIAN PHOTO).
Left: Rev. James (J. J.) Tompkins (REID'S PHOTO CENTRE).

their family farms in favour of industrial jobs in local centres or else-where in North America. To the people of this region, "Upper Canada" seemed to be the federal government's "centre of the universe."

Most of the people that settled the Cape Breton Highlands were of Scottish decent with strong ties to the Roman Catholic Church. The settlers worked the land, established small family farms, and like the independent fishers along the coastline of Cape Breton and mainland Nova Scotia, were also poverty-stricken and had very little to show for their efforts. They were a hard-working and industrious lot, but like the shore fishers, most were illiterate or had very little formal education. The village priests were looked upon by the local folk as being very educated and learned men, and as such were held in high regard. They were relied upon to not only provide the neces-sary spiritual and moral leadership to help their parishioners through difficult times, but they also became the prime source of support for land and legal issues. Sometimes the simple task of writing a letter fell upon the priest. Because of their standing within the community, the priests were in a position to express the concerns of the people to the Catholic diocese that was headquartered in Antigonish, NS.

The Antigonish Movement that had its beginnings in the 1920s with Moses Coady and Jimmy Tompkins arose out of a need to develop solutions to the problems that had plagued the people of eastern Nova Scotia since the late 1880s. The migration of young farmers, fish-ers, and other workers to jobs in the growing manufacturing sector of Central Canada had seriously eroded the region's voice in federal politics. The Maritime Rights Movement of the 1920s had failed, and the provincial government tended to support business interests rather than address the growing concerns of the worker. The common worker and fisher began to resent the provincial government, and any trust held for this democratic institution was slowly eroding.

Co-operative marketing organizations began to appear in British North America in the 1840s. At that time, labour attempted unsuc-cessfully to start stores common to those in Britain, with the first one developed in 1861 in the mining town of Stellarton, NS. Others started to appear in various parts of Nova Scotia but soon closed as a result of the recession that followed the First World War and the fact that those who had membership in such ventures did not have the knowledge nor the education necessary to be able to succeed. More important, each of the co-operatives operated independently with no consolidation of effort with regards to the management of markets and the provision of goods and services. The reasons behind these failures did not go unnoticed by Coady and Tompkins, who firmly believed that if the people were to move forward then they themselves must be provided the education and social tools to resolve local problems. They therefore based their philosophies on the following tenets:

- Primacy of the individual.
- Social reform must come through education.
- Education must begin with economic reform.
- Education must be through group action.
- Effective social reform involves fundamental changes in social and economic institutions.
- The ultimate objective is a full and abundant life for everyone in the community.

In the early 1920s, Coady and Tompkins were located at St. Francis Xavier University (StFX) in Antigonish, NS. At this time, StFX consisted of a high school and a university. Rev. Coady was the principal of the high school while his older cousin, Rev. Jimmy Tompkins, was vice-president of the university. Both believed that people could collectively solve their economic problems through a program of adult education and by working collectively through "co-operation" rather than as individuals.

Being brought up in rural Cape Breton, Coady and Tompkins knew well the poverty and problems being experienced by the Cape Breton farmers. Both had studied abroad and saw first-hand how people in other countries overcame their social and economic problems through a system of education, social reform, and co-operation. Both were very vocal in their belief that the universities were failing the common worker and that they only provided education for the rich. Thus, the common worker, regardless of intellect, could never gain a formal education. In 1921 Tompkins started the "People's School" to demonstrate that adult education had the power to change one's thinking and solve collective problems. Both men had a fervent belief that the people must collectively find solutions to their problems by themselves and not rely on government to force a solution on them. It was often said that Father Tompkins wouldn't even let your cat out for you, but he would stay up half the night nagging you to do it yourself. Drs. Coady and Tompkins continued to lecture their fellow parish priests on the need for collective co-operation and education.

In 1921 and 1922, Father Tompkins took the lead in a very ambitious and controversial effort to create a single university that would be known as the University of Nova Scotia. He believed that the sectarian nature of the universities in Nova Scotia and eastern New Brunswick was failing to prepare their graduates for the changes that the postwar world was presenting. This notion was slowly gaining the support of the local and the provincial media, foremost among them being the Halifax *Herald*. The editorial on May 28, 1921, said the following: "For many years universities everywhere have been, and most of them still are, labouring under the misconception that they, by divine right, shall serve the wealthy and privileged classes, and in no degree promote popular education."

Tompkins's ideas and philosophies did not sit well with church leaders of the various denominations in Nova Scotia and were definitely not well received by his own masters at StFX. In 1922, Father Jimmy Tompkins was removed from his position and "exiled" to Canso, a fogbound fishing village on the eastern tip of Nova Scotia. No one at the time could have predicted the impact this seemingly "punitive exile" was about to have on the maritime fishery and the history of Nova Scotia. Father Jimmy Tompkins was about to become the spiritual leader and father of the co-operative movement in the Nova Scotia shore fishery.

On a cold and bitter morning in December 1922, Father Jimmy Tompkins arrived at his new charge in Canso. Within his area of responsibility were three churches: the Star of the Sea Church in Canso; St. Agnes Church in the adjoining village of Little Dover; and St. Vincent de Paul Church in Queensport, a very small fishing village bordering Chedabucto Bay and some twenty kilometres distant from Canso. A quick scan of his new charge was less than encouraging. The area was one of utter desolation. Poverty and malnutrition were widespread throughout his parish, and worst of all, evidence of the lack of concern for the economic plight of the inshore fishers was everywhere. Even his rectory was in total disrepair with little heat and so drafty that he once commented, "a man could get blown off his feet in the corridor."[6] Still stinging and bitter over the treatment he received in Antigonish, "Father Jimmy," as the citizens of the area would come to know him, was rather dejected and bewildered. However, like a good soldier, he accepted the "orders from his commander" and vowed to himself and God that he would serve the people of his parish to the best of his ability.

Instead of bemoaning his fate, Father Tompkins soon realized that Canso had the right conditions for him to implement the policies and changes that he and Coady had been advocating. Within a very short time, Father Tompkins gained a full appreciation and understanding of the problems facing the local fishers and initiated a program of "study groups" that he believed would allow the people to develop solutions. One of the first priorities of the study groups was to put in place a program of adult education. His belief was that with his assistance, the people could start to educate themselves, start looking at the deplorable conditions in which they lived, and moreover, start discussing how they could improve their outlook on life and regain their dignity. He became a strong advocate for the formation of co-operatives and credit unions and firmly believed that establishing such facilities in Canso and the surrounding area would bring life and prosperity back to the region. He became quite vocal in his belief that co-operation, not individualism, would be the key to recovery. Throughout this time, he remained in close contact with Coady and his colleagues in Antigonish and the rest of the dioceses throughout the province. They began to watch with interest Father Tompkins's work with the fishers in Canso and believed the time was right for change, if only there was something that could bring the people together.

CHAPTER 3

THE CANSO INCIDENT

*If the masses of people have become, in a
sense, slaves it is because they have not
taken the steps or expended the effort
necessary to change society.*

– MOSES M. COADY, *MASTERS OF THEIR OWN DESTINY: THE STORY
OF THE ANTIGONISH MOVEMENT OF ADULT EDUCATION THROUGH
ECONOMIC COOPERATION*

THE YEAR 1927 WAS A SPECIAL ONE FOR THE YOUNG DOMINION OF
Canada. It marked the sixtieth anniversary of Confederation, and most
communities across Canada were caught up in the festive atmosphere
of the pending Diamond Jubilee celebrations scheduled for July 1. The
federal and provincial governments established Jubilee Planning
Committees that planned gala events for that date, and the lists of
dignitaries seemed never-ending. The citizens of Nova Scotia, like
the rest of the nation, were caught up in the preparations for their
own community events. The Halifax *Herald* contained daily reports
of the plans being made in every community across the province. It
was reported that Canso was planning a "striking ceremony." Local
businesses and buildings were being decorated, as were vehicles.
Gala parades were planned that involved the citizens and especially
schoolchildren. Considerable publicity was given to the fact that the
speeches by the mayor and his special guests, William Duff, MP, and H.
H. Rice, MMP, would be highlights of the festive proceedings and that
the festivities would conclude with an elaborate display of fireworks.
There is no question that the events that were to transpire in Canso

on July 1, 1927, would be most memorable indeed. The fallout of what happened that day in Canso would have a major effect on the Nova Scotia fishery and the economics of the province for decades to come.

The morning of July 1 broke bright and fair. Most of the fishers remained ashore. Their idea was that one might just as well enjoy something, as going out to fish would not be worth the effort. The daily program of celebration was carried out as planned. Father Tompkins was taking in the events when, walking on the docks, he happened on a forlorn fisher. When the priest asked him why he was not taking in the events, the fisher answered that there was nothing to be cheerful about. The fisher then vented his frustration on Father Tompkins, explaining the problems that existed in the area and the difficulties the current conditions were having on the fishers' ability to survive let alone turn a profit or pay down their debt to the local fish buyers. By this time other fishers had joined the conversation, and instead of just venting and going home, with Father Tompkins's support they staged an open protest of sorts that could not be ignored. Some fishers attended the festivities with the Union Jack and Canadian Ensign flags displayed upside down. They became somewhat vocal about their situation and soon had the ear of the reporters that were present. Seizing on the moment, Father Tompkins took it upon himself to organize a meeting with the fishers in a local hall that evening.

There have been many accounts of what happened in Canso that day, but perhaps the best was the article that appeared in the Halifax *Chronicle* on July 7, 1927.

The Dominion Jubilee Celebrations at Canso took a peculiar turn when in the evening, some forty fishermen met at a local hall and after a heated three-hour discussion, unanimously passed a resolution to call upon William Duff, MP and Hon. John A. Walker to come to Canso at the earliest possible moment to discuss and deal with the pressing problems of the local fishermen.

The fishermen say that they feel they have been discarded and trifled with in the past and that it rests with the govern-ment—provincial and federal—to take the situation in hand and check conditions that they say are fast reducing the fish-ermen and their families to starvation.

The fishermen see no reason why the government shouldn't use some of the money now expended on immigration, to provide means of subsistence for Canadians who are already in the country.

One fisherman taking part in the discussion stated that his return from employment with one of the large local fisheries allowed him only ten cents per meal for each member of his family with no allowance for clothing and other expenses in connection with the household. His condition is said to be, in a measure, illustrative of what is general among the fishermen in the vicinity.

The methods of the Wheat Pool of Western Canada are making a strong appeal to the Canso fishermen and they contend that it is high time that up-to-date methods and modern machinery be installed to save the people engaged in the fishing industry, and the industry itself for Canada. Proper freezing facilities and organization of the fishermen are crying needs.

This heated discussion and the anger demonstrated by the fishers in Canso on July 1, 1927, prompted another meeting later in the month to discuss the grievances of the fishers more thoroughly. This meeting was covered by the press, and without holding anything back, reporters provided to their readers and the population at large vivid details of the gathering that was attended by the Honourable John A. Walker, minister of natural resources, and William Duff, MP for the region. This meeting resulted in a number of draft resolutions relative to the lack of a proper market; the selection process used by the dealers when buying fish; the incursions the new steam trawlers, both Canadian and foreign, were making into local fishing grounds; and the effects of the migration of people from Canso. But perhaps the most important resolution was a request to the federal government in Ottawa to convene a Royal Commission to investigate all phases of the fishing industry and to make recommendations as to how this industry could be sustained for future generations.

Father Tompkins, having the full trust and confidence of the local fishers, kept up the rhetoric with his fellow priests in the Antigonish dioceses and gained similar support from a number of Protestant clergy along the breadth of the southern shore of Cape Breton and the Eastern Shore of Nova Scotia. That same July, a general meeting of all priests in the dioceses was held at the campus of St. Francis Xavier University in Antigonish. Tompkins and Coady brought the situation with the Eastern Shore fishers to the floor of the meeting and ensured clergy present were made well aware of the situation and how they could assist. Tompkins went so far as to hold a special meeting of all the priests (and some Protestant clergy) from the fishing villages in the dioceses. At the end of their deliberations, they sent telegrams to officials both in Halifax and Ottawa describing the unsatisfactory and deplorable condition of the fishery and demanded that the situation be investigated. Under normal circumstances, making such demands of the federal and provincial governments would have had no impact, but the ideas and work of Drs. Coady and Tompkins were gaining traction and had the full support of the local and national media.

Perhaps one of the most avid supporters of the fishers was Mr. George Farquhar, the then editor of the Halifax *Chronicle*. From the time of the first meeting in Canso on July 1, 1927, Mr. Farquhar penned and published a series of seven articles called "Save the Fishermen." These articles provided a very graphic description of the conditions that existed among the fishing communities along the Eastern Shore of Nova Scotia. His articles soon caught the interest of the reading public and brought added pressure to bear on the politicians within the provincial and federal governments. It soon became blatantly clear that the situation in the Nova Scotia fisheries could no longer be overlooked and action had to be taken immediately.

On August 5, 1927, the federal minster of fisheries and marine, the Honourable Arthur Cardin, announced that a Royal Commission would be established to investigate all phases of the fishing industry of the Maritime provinces.

CHAPTER 4

THE ROYAL
COMMISSIONS

*The more his equipment confines him to
the inshore fishing, the more will his catch
depend on the inshore runs of certain species,
most of which will apt to be seasonal.*

- FROM THE SUMMARY OF THE *REPORT OF THE ROYAL COMMISSION
ON PROVINCIAL DEVELOPMENT AND REHABILITATION*, 1944

IT HAS BEEN SAID THAT IF YOU WANT THE GOVERNMENT TO APPEAR
that they are concerned about an issue and are working hard to solve
the problem but then end up doing nothing, have them form a Royal
Commission. Between 1919 and 1945, there were no less than seven
separate Royal Commissions that had as part of their mandate an
examination of some facet of the Maritime fishing industry or the
economics thereof. They are as follows:

- Royal Commission on Maritime Claims (1926)
- Royal Commission Investigating the Fisheries of the Maritime
  Provinces and the Magdalen Islands (1927)
- Royal Commission, Provincial Economic Inquiry (1934)
- Royal Commission on Price Spreads (1935)
- Royal Commission on Financial Arrangements between the
  Dominion and Maritime Provinces (1935)
- Royal Commission on Dominion–Provincial Relations (1940)
- Royal Commission on Provincial Development and Rehabilitation
  (1944)

It is not surprising given the deplorable state of the inshore fishery that most of the commissions identified major shortcomings in the Maritime fishing industry and all made recommendations relative to the topic under investigation. However, of the seven, only three were very critical of the lack of provincial and federal government support to the Maritime fishery and proved to have any positive impact on the Nova Scotia fishery. Of those three, only two had any substantial long-term positive impact. It was the Royal Commission Investigating the Fisheries of the Maritime Provinces and Magdalen Islands (1927), also known as the MacLean Commission, and the Royal Commission on Provincial Development and Rehabilitation (1944)—the portion of the final report dealing with the fisheries became known as the Dawson Report—that not only tabled specific recommendations, but put in place a process to ensure the recommendations were reviewed by the highest authorities in both the provincial and federal governments and held both levels of government accountable to ensure progress was being made. These two commissions can be credited with forcing both levels of government to recognize the need for formal intervention and support to the Maritime fishery and in particular the inshore fishing industry of Nova Scotia.

ROYAL COMMISSION INVESTIGATING THE FISHERIES OF THE MARITIME PROVINCES AND THE MAGDALEN ISLANDS (1927)

Shortly after the formal announcement of its formation on August 5, 1927, the commission got down to work. The Hon. Mr. Justice A. K. MacLean chaired the commission, hence the "MacLean Commission." This body was instructed to investigate the fishing industry in Nova Scotia, New Brunswick, Prince Edward Island, the Magdalen Islands, as well as some of the coastal portions of Quebec that bordered on the Maritime provinces. The mandate was very clear, and the direction given to its members was to inquire and provide recommendations on the following matters:

- What should be done to increase the demand for fish both at home and in foreign markets?

- What is the spread of the price of fish between the producer and the consumers? If considered excessive, what should be done to remedy the condition?
- What should be done to develop the inshore fisheries to their full capacity?
- Should there be further restrictions on steam trawlers operating from the Canadian Atlantic ports, and if so, what should they be?
- Keeping in mind that no exceptional privileges were available for McCain fishing vessels visiting from United States ports, should the so-called modus vivendi privileges or any of them be renewed? [7]
- Should the amount now annually distributed as fishing bounty be continued on the present basis? (For an explanation of "bounty," see the sidebar in chapter 7.)
- Should there be an inspection of fresh fish of all kinds landed, placed in storage, and shipped from the coastal points?
- Should there be an inspection and grading of dried fish?
- Should there be any modifications in the present laws and regulations of the fisheries?

There is no question that the terms of reference charged to the commission were ambitious. However, during its initial deliberations, it became very clear that there were major problems within the shore fishery that were outside the scope of the commission's original mandate. The Hon. Justice MacLean realized that if the commission were to have any credence with the fishing industry, then all issues must be investigated and addressed. He therefore decided to investigate all facets of the Maritime fisheries and concentrate on the shore fishery where he rightly believed the major problems existed.

At the hearings held throughout the Maritime provinces and the Magdalen Islands, scores of local fishers were allowed to testify, but only a handful of offshore fishing captains and large fish dealers and processors chose to attend. It was the commission's consensus that the large offshore fleet and their supporting processors were exacerbating and masking the problems with the shore fishery. It appeared that the larger fish processors had more influence with the federal and provincial governments and were benefactors of any governmental funding at the expense of the shore fishery. In reviewing the evidence presented, the Hon. Justice MacLean came to believe that although there were some issues with the offshore or banks fishery, most of them were in line with the original terms of reference and could be addressed within that scope. However, the problems that were being uncovered in the shore fishery required a deeper investigation, full disclosure, and substantive recommendations that were perhaps never considered when the commission was established.

In September 1927, Rev. Coady made representation before the commission at its meetings in Halifax. In his address, he held nothing back in discussing and outlining the current state of affairs in the small coastal communities of Cape Breton and the Eastern Shore of Nova Scotia. Coady believed that similar conditions existed in the majority of small fishing communities throughout the Maritimes and it was time for governments at all levels to start listening to the concerns of the fishers. He also outlined the need for education and his idea that solutions could be found if the fishers worked together collectively. So strong was his representation at the hearings, the Hon. Justice MacLean brought the issue of education and co-operative organization to the forefront of the commission. MacLean and the members of the commission were very intrigued by and impressed with Coady's ideas about education for the fishers, the establishment of a co-operative marketing system, and providing the tools, in the form of better fishing craft and methods, that would allow Nova Scotia fishers the ability to determine their own destiny.

Over the course of its deliberations the MacLean Commission held forty-nine hearings in New Brunswick, Prince Edward Island, Nova Scotia, and the Magdalen Islands, most of which extended over several days and collectively had 823 persons appear to present evidence. Upon conclusion of the formal deliberations, the MacLean Commission filed its formal report consisting of over 5,700 pages in which a number of dramatic recommendations were made. There was no question that the original terms of reference were addressed; however, the findings and recommendations relative to the shore fishery were most candid and startling. Some of the bolder ones included the following:

- Restrictions were to be placed on foreign steam trawlers operating from Canadian ports and in Canadian waters, the majority of which came from the United States. The commission recommended that foreign steam trawlers be no longer allowed to fish in Canadian waters and that any fish landed by a steam trawler in a Canadian port would be subject to a tariff of one-third to one cent per pound. Steam trawlers operating out of Nova Scotia ports had to be Canadian-owned and registered in a Canadian port.[8] Such trawlers were required to be licensed, and in Nova Scotia the number of licences was restricted to three.

• Fishers should be assisted to organize co-operatives with a qualified organizer appointed to carry out the work.

• Education among fishers was to be promoted, similar to the education for farmers.

• A formal system for the inspection and grading of fish products should be introduced and enforced.

In addition to the recommendations, the MacLean Commission was very critical of the lack of provincial and federal government funding or support of any kind to the Maritime fishery. Millions of dollars were being given to private entrepreneurs for the expansion of the manufacturing sectors, but nothing in the form of either loans or grants was being provided to the Maritime shore fishers. The report was equally critical of the lack of responsibility being shown by the federal government in addressing the existing problems in the Canadian fisheries as a whole. In his personal comments, the Hon. Justice MacLean strongly emphasized the importance of the fishing industry to the economic growth of Canada and that it should not be underestimated. He also made it very clear that there should be a separate ministry established in the Canadian Parliament with sole responsibility for fisheries. If the inshore fishery was to move forward, government at all levels had to become engaged.

The events that occurred in Canso, NS, in July 1927 and prompted the Royal Commission were still fresh in the minds and pens of the local and national media. It is not surprising that most major newspapers in Canada and even some from the United States closely watched the deliberations of the Royal Commission. When the findings and the recommendations of the commission were released, it was the media that kept the local population informed and ensured that the politicians at both levels of government acted upon the recommendations.

Change did not occur overnight, but it was not far on the horizon. In August 1929 the minister for the Department of Marine and Fisheries, the Honourable J. A. Cardin, recommended that Rev. Moses Coady be appointed to organize co-operatives among fishers.

The minister was most impressed with Coady's representation to the commission during the Halifax meetings and believed he had the necessary experience and knowledge and the trust of the fishers to be able to successfully implement the recommendations he had made to the commission. In 1930 the federal government established the Department of Fisheries as a standalone ministry and passed legislation that imposed restrictions on steam trawlers as recommended by the MacLean Commission. Within a few years, a proper fish-grading system was developed and implemented, but unfortunately the much-needed funding for better vessels and fishing gear had yet to be realized.

THE ROYAL COMMISSION ON PROVINCIAL DEVELOPMENT AND REHABILITATION

On June 25, 1942, the Government of Nova Scotia appointed a Cabinet Committee to deal with the rehabilitation and development of all aspects of the Nova Scotia economy after the end of the Second World War. On May 12, 1943, following a recommendation of the committee, a Royal Commission was created. The Crown appointed Dr. Robert MacGregor Dawson to head this effort and act as commissioner for the proceedings. Dawson was a noted and respected political scientist who at the time of his appointment was teaching at the University of Toronto.

The purpose of the Royal Commission was to investigate, survey, consider, and report on measures to aid the rehabilitation of those discharged from the Armed Forces and from war industries at the end of hostilities and to make recommendations on how the economy of Nova Scotia could transition from a wartime footing and grow in peacetime. Every facet of Nova Scotia industry and economic development was to be investigated and studied. Mr. Stewart Bates, professor of Commerce at Dalhousie University, Halifax, NS, was chosen to head up the investigation into the Maritime fishery.

Upon accepting his appointment, Mr. Bates chose not to confine the investigation solely to the issue of returning service persons and those that would be displaced by the downsizing of war-related industries, but rather to conduct a complete examination of the problems inherent with the total fishery and determine what impact, if any, the ending of hostilities would have. He also felt it important to measure the progress that was being made since the release of the MacLean Commission report of 1927. His approach, dedication to the task at hand, candid findings, and recommendations resulted in the development of a report and follow-up process that became one of the most important documents in the history of the Nova Scotia fishery.

During his investigation and subsequent report, Mr. Bates made some very startling observations. His research revealed that 64 per-cent of the total catch volume of groundfish and flatfish taken in 1939 in Nova Scotia was landed at the ports of Halifax, Lunenburg, North Sydney, and Lockeport. What was more startling is that the catch landed at these ports equated to an average of 80,000 pounds for every person that went to sea in the larger offshore vessels operat-ing from the aforementioned ports, while in the shore fishery the total catch, which included lobster, herring, and mackerel, averaged 12,500 pounds. His investigation further revealed that the catch fig-ures used by the federal government were not broken down by the area fished or whether the catch was gained by inshore or the off-shore fishery but were treated as a single figure. Therefore, there was a false impression that the Nova Scotia fishery was indeed doing well and growing. Had the catch figures been properly broken out and analyzed, they would show that 10 percent of the fishers operating from four ports in Nova Scotia landed over two-thirds of all flatfish and groundfish species taken. The reasons for this disparity were the fact that the shore fishers were very poorly equipped, had very small craft that allowed them only eighty to one hundred days of fishing per year, and that most small ports lacked the wharf and fish-handling capacity to compete with the larger centres.[9] What was being ignored was that the aforementioned ports were the main operating centres for the large offshore fleets that were better equipped, had larger catches per capita, and fished a longer fishing season.

Bates's findings were extremely critical of both levels of govern-ment for their seeming lack of concern for the plight of the shore fisher while at the same time planning an expansion that favoured the larger fish processing plants and fleets. He made it very clear that the catch made by a fisher in any given period of time depends on the equipment, the efforts, and the availability of fish. Without adequate equipment and the support to obtain such equipment, most shore fishers would be confined to inshore runs of certain species that are more apt to be seasonal. In his summation, Mr. Bates stated that unless there were changes in governmental policies to provide support to the shore fishery at both the federal and provincial level, any hope of expansion would be nonexistent.

The findings of the commission made it equally clear that the shortage of capital and the long depression retarded the development of the Nova Scotia fishery. In addition, the commission was equally critical of the deep-rooted attitude of the fishers and processors to dwell on the traditional methods of fishing—such as the use of sail-fitted fishing schooners and dories, most of which were becoming obsolete or no longer profitable to sustain—and the unwillingness across the industry to experiment with new methods of fishing and processing. Mr. Bates did acknowledge that while the conditions that existed in most of the small fishing communities in Nova Scotia before the war were out of the control of the federal government, that same government was negligent and not assuming its full constitutional responsibility for the fishing industry.[10] Although the federal government had provided some limited support with respect to world markets and trade policies, any gains made were erased by national trade policies and tariffs that existed among the provinces. As a result, the government knowingly or not knowingly had placed undue hardship on the Nova Scotia fishery and weakened the ability of the fishers to become competitive within their own national borders, let alone on the world stage.

With respect to the returning servicemen and -women and those that had been involved in the war industry, the commission went to great lengths to explain in detail some of the financial benefits they would receive as a result of their service. However, it was made perfectly clear that the rehabilitation of war veterans was entirely the responsibility of the "Dominion government." Conversely, it was the responsibility of the commission to expound on the concerns of the provincial government of Nova Scotia for those returning from the war and associated war industries, to ensure there would be ways of earning a livelihood within the province. In examining the Nova Scotia fishery relative to these benefits, Mr. Bates explained, the benefits available to those who served or who were eligible as a result of their involvement in war-related industries were a starting point for those that wished to return to their community and re-enter the industry. But he made it very clear such benefits should not be factored by both levels of government as part of the funding equation for the fishery as a whole. Such benefits would not apply to those that remained in the fishery and supported the war effort in this fashion. Any consideration of such benefits in determining the approval of a financial support envelope would place non-veterans at a severe disadvantage when it came to acquiring better equipment and vessels. Therefore, it was imperative that federal or provincial funding that had been earmarked for the expansion and modernization of the fishery be applied to individuals in the industry as a whole and not to specific individuals. In his final report he emphasized that more tangible and realistic financial support from the federal government was required.

Mr. Bates was equally critical of the financing approaches taken or being contemplated by the provincial government. In his view, the Nova Scotia Fisherman's Loan Board, established in 1936, was in itself seriously flawed and ineffective. He concluded that its funding envelope was incapable of providing for the number of fishers requiring financial support and of granting loans sufficient for the modernization of their boats and gear. In the view of the commission, the way the loan board conducted its business was no different than fishers trying to gain funding for the acquisition of new boats and gear from local merchants. On the federal side, the commission was extremely critical of the government allowing the West Coast fishery singular access to a subsidy known as the Federal Fishing Vessel Construction Assistance Program for the construction of new fishing vessels or the conversion of existing vessels to another fishery and which provided a number of substantial tax benefits for those who took advantage of the program. Although the program was extended to the Maritime fishery in July 1942, it only applied to the construction of trawlers/draggers and/or conversion of schooners to facilitate the use of otter trawl or longlining. Notwithstanding the need to increase fish production in support of the war effort, Mr. Bates stated that this program in its current form placed the Nova Scotia fishery at a distinct disadvantage and displayed inequity and favouritism. The commission strongly recommended that a complete overhaul of the Nova Scotia Fishermen's Loan Board practices and policies be carried out without delay and that the subsidy being granted to the West Coast fishery be applied equally to the Nova Scotia fishery.

With respect to the findings and recommendations tabled by the MacLean Commission in 1927, Mr. Bates acknowledged that some progress was being made but that more concrete progress would have been made if the federal government had carried out its constitutional duties to the fishery. He acknowledged that the effects of the Depression and the need to concentrate on the war effort did affect the government's response to certain facets of the Canadian fishery; however, such events should not have resulted in a complete neglect of the Nova Scotia fishery. He was very quick to acknowledge the fine work being done by Coady in the formation of co-operatives for fishers. The commission fully endorsed the need for adult education among the fishers and collectively working together to solve common problems. It was the finding of the commission that there had been some improvement in this area since the release of the MacLean Commission report in 1927; however, there was still a lot the province could do to facilitate the process. Mr. Bates stressed that in the coming years, the shore fishers needed to have a better knowledge of how the Nova Scotia fishery was operated with respect to processing, quality control, and basic market strategy if they were to maintain and grow their competitiveness. He also cautioned that the limited funding that had been provided to date by the federal government for the building of new co-operative–sponsored facilities required a better distribution mechanism if the industry was to thrive in the future. Co-operative ventures did not exist in every fishing port of Nova Scotia; therefore, there should be no discrimination in providing government aid to either the co-operatives or to private firms. If the Nova Scotia fishery was to exist and expand, then governments at all levels needed to assist both fishers and processors along the full breadth of the Nova Scotia coastline. Mr. Bates was quick to point out that the Nova Scotia and the Maritime fishery as a whole were lagging behind in methods, vessels, equipment, processing facilities, and a fully supported marketing strategy when compared to the fishery being conducted along the Eastern Seaboard of the United States and along Canada's own West Coast. The commission called on both levels of government, federal and provincial, to invest in the research and development of better vessels particularly for the shore fishery, better fishing equipment and methods, refrigeration facilities, and processing of catch.

The commission also addressed the issue of the Canadian trawler restrictions. In reporting his findings on this matter, Mr. Bates stated that if there was to be any expansion of the Nova Scotia fishery and economic growth in this sector, then those involved in the fishery had to recognize the trawler as a technological advancement in fishing methods. Banning the construction and the operation of trawlers from Canadian ports while those from foreign countries, primarily the United States, operated on fishing grounds within three nautical miles of the Nova Scotia coastline did not make economic sense. It was his recommendation, and that of the commission, that this ban be lifted and trawlers be allowed to operate from Canadian ports as part of the whole strategic plan for fishery expansion.

CHAPTER 5

THE CO-OPERATIVES

*Today we are packing and shipping our own
lobsters to foreign countries through our
co-operatives, and we are processing our
own fish. We have our own icehouse and our
own grocery store all through organization,
co-operation, and the sale of one crate of
lobsters!*

– CAPT. BILLY TOM FELTMATE, 1937

THE FINDINGS OF THE ROYAL COMMISSION INVESTIGATING THE
Fisheries of the Maritime Provinces and the Magdalen Islands, known
as the MacLean Commission, were completed and presented in June
1928. It was not until August 1929 that the recommendation regarding the organization of the Maritime fishers and the formation of
fishermen's co-operatives was finally addressed. The formation of
those co-operatives did not sit well with the larger fish processors
and the independent buyers. For the first time in the history of the
Nova Scotia fishery, there would be an accounting, and the "scalping"
of the shore fishers would come to an end. The recognition of the
need for the medium-sized longliner was still some twenty years in
the future, but the seeds so needed for the shore fishers to develop
and possess such a tool were finally being sowed.

Coady's brief to the commission had such a profound impact on
the proceedings that the commissioners unanimously agreed that
if there was any one individual that could accomplish the daunting

The Guysborough Fishermen's Co-operative plant in Whitehead, NS, originally opened in 1946, as it stands today. It was once supported by three Cape Island-type longliners and a number of smaller fishing boats and employed about twenty people. It processed groundfish for the salt and fresh fish markets, had a medium-sized cold storage facility, and packed and shipped wild blueberries in season. Today it is used as a lobster pound. (AUTHOR'S COLLECTION)

task of organizing the shore fishers, it was Coady. The Honourable J. A. Cardin of the Department of Marine and Fisheries asked Coady if he would act on behalf of the department and undertake the task of organizing the Maritime fishers. Coady accepted the task with the fervent belief that through education and working together, the fishers collectively could influence the market beyond their local area. He, along with his cousin Tompkins, had recognized for some time that the shore fishers needed better boats and equipment if they were to succeed and better their lot. But most of all they had to be somehow transformed from hard-working, uneducated fishers into masters of their own destiny, capable of competing with the larger fish dealers and processors. The answer lay in the successful organization of the fishers and the formation of formal co-operative businesses.

Coady also knew that establishing separate co-operatives in the small fishing communities along the Nova Scotia coast would be of little value unless the there was a federation of fishers across not only Nova Scotia, but also Prince Edward Island, New Brunswick, and the Magdalen Islands, and that they remain united in matters regarding the marketing of their product and the protection of the fishers themselves. He felt quite confident that with the support of his fellow priests and the staff at StFX University he could achieve success.

In accepting this assignment, Coady was well aware that the road ahead would be very difficult. The successful organization of the Maritime inshore fishers and the implementation of the co-operative movement could only be achieved if first a number of economic and social obstacles were identified and rectified. There was also a need for formal legislation within each of the Maritime provinces that would allow the incorporation of the fishermen's co-operatives so they could be recognized and authorized to act as legal bodies. In Nova Scotia the latter was facilitated by existing legislation passed in 1916 that provided for the incorporation of fishermen's co-operative societies. In 1927 another act was passed that allowed for the organization of fishermen's federations. Similar legislation existed in New Brunswick. In 1935, in response to recommendations made by the Royal Commission on Provincial Economic Inquiry, the Co-operative Associations Act was passed in Nova Scotia under which all classes of co-operatives could be incorporated. Similar legislation was passed in Prince Edward Island.

With respect to the fishers themselves, there was a quick realization that a number would be very reluctant to abandon their current fish buyer for a concept that, although very attractive, in their eyes had yet to be proven. In some instances, the local fish buyers had already threatened the fishers by stating that they would no longer support their needs or purchase their catch if they joined the co-operative movement. In areas of Nova Scotia and the remaining Maritime provinces, there were some issues with religious sectarianism. It was not uncommon in the Nova Scotia of the 1920s and 1930s to have a clear divide between Protestants and Catholics, even in small communities along the coast. Given that the idea of the co-operatives originated from within the Catholic Church, it was seen by some Protestant-dominated communities as an incursion by the Catholics into their way of life. If the local dealer was a Protestant, there was a tendency to remain loyal to that individual regardless of his business scruples.

Another obstacle was the lack of any concrete funding from any level of government or the availability of any financial incentives that could be used as start-up funding for the project. Unfortunately, the only funding Coady received from the federal government was a sum of $5,000 for set-up and travel expenses. Any funding and expenses related to the formation of co-operatives would have to be borne by the fishers themselves, and they were already struggling to survive financially.

In keeping with the terms of reference for the MacLean Commission, Coady's assigned task encompassed Nova Scotia, New Brunswick, Prince Edward Island, and the Magdalen Islands. Although there was some commonality with respect to the plight of the fishers in each of these areas, there was diversity with respect to fishing methods, species, and markets. Coady contended that if the fishers had any chance of establishing a co-operative movement, regional differences must be set aside. He also realized that he had to convince the fishers not to fear any threats or retribution that could come from the existing dealers should they join a co-operative movement.

All the while that Coady was preparing for his work with and for the MacLean Commission, his fellow priests along the Eastern Shore were continuing their work and study groups with the local fishers. In the community of Larry's River, the local priest, Father Forest, organized citizens and those from the adjoining community of Charlo's Cove to build a new two-room school. This project demonstrated to the villagers what was possible if they collectively worked together for a common goal. In the coming years, Father Forest would play a major role in the formation of a number of successful fishermen's co-operatives along the Eastern Shore of Nova Scotia, including the large fish plant at Port Bickerton.

Of all the priests and clergy involved in the organization of the fishers, foremost among them was Father Jimmy Tompkins. It was not uncommon for Father Jimmy to travel from Canso to the surrounding small fishing villages like Little Dover, Whitehead, and Port Felix. Instead of preaching on religious doctrine, he would invite fishers to talk about their problems and see if there was something that could be done. It was not long before Father Jimmy broke down secular boundaries and gained the confidence of the fishers. His presence was welcomed wherever he went. The stage was slowly being set to form the first co-operative.

Like Coady, Father Jimmy knew there had to be some concrete demonstration to the fishers that in forming a co-operative they would better their lot and be able to control their future. In 1928, and just prior to the release of the findings of the MacLean Commission, at Father Jimmy's urging, Captain William Feltmate (known as Captain Billy Tom to his friends) and a few fishers from the village of Whitehead, NS, came together to ship 140 pounds of lobster direct to a buyer in Boston. The local dealer was paying seven cents per pound. Had they sold to him, the 140-pound crate would have been worth $9.80; divided among the four fishers, that would leave each man $2.80. Instead, the fishers received a cheque from the Boston dealer for $32.00. The four fishers received $8.00 each for their share of the shipment. The price they received was four times what the local dealer was paying. The success of this venture convinced the fishers of the merits of the co-operative movement, and a few months later the fishers formed the Whitehead Co-operative Society. The first fishermen's co-operative in Nova Scotia had been formed. This small co-operative had very humble beginnings, but it was a start. Initially there were approximately eleven shareholders; over time the co-operative grew. With the help of Father Jimmy, they were able to use their co-operative to buy fishing supplies such as twine, wooden laths for lobster traps, and even gasoline at far better prices than they were receiving from local dealers. In December of 1929, the idea of the co-operative started to expand to the neighbouring Acadian community of Port Felix. With the help of the local priest, the sectarian wall that existed between the predominately Acadian Catholic community of Port Felix and the Protestant village of Whitehead started to crumble, and the interaction between the two villages became a model for the co-operative movement. Captain Billy Tom Feltmate could not have put it any better when at a co-operative meeting, he stated, "Over in Port Felix and previous to our organization there was a line fence, so to speak, between us, and each fellow kept on his own side of it, Protestant on one side and Catholic on the other. We had no business transactions in any way. Since we fishermen began to organize and co-operate, we have become fast friends."[11]

It did not take long for the news of the lobster sale in Whitehead to get around. Not long after this incident, Father Tompkins brought forty fishers together from Dover—a small, isolated, and poverty-stricken village halfway along the coast between Whitehead and

Above: Captain William (Billy Tom) Feltmate and his wife, Margaret.
(PEGGY FELTMATE)

Left: The first Fishermen's Cooperative, Whitehead, NS.
(PEGGY FELTMATE)

A gathering of people at Whitehead, NS; they had been invited by Father Tompkins and were interested in learning more about the Co-operative initiative. (PEGGY FELTMATE)

Canso—to form the Dover Lobster Co-operative. Father Tompkins personally lent the group $300 and arranged a loan for an additional $700. With their own sweat, blood, and blind faith they built a small lobster factory that could process in-season and ship lobster direct to a buyer again arranged by Father Tompkins and some of his associates. By 1934, this small enterprise had turned over $10,000, a feat that a few years before would have been thought to be impossible. These two small co-operative initiatives played a significant role in helping Coady's work in organizing the fishers.

Captain Billy Tom Feltmate became a staunch supporter of the co-operative movement and the merits of the fishers working together. Again, it did not take long for the story about the formation of a small co-operative in Whitehead and the formation of the processing plant in Dover to reach not only the ports along the Eastern and South Shores of Nova Scotia, but the other Maritime provinces and the Eastern United States. Capt. Billy Tom was often asked to speak about this experience at various gatherings sponsored by

Father Tompkins. It was not long before local newspapers and provincial papers such as the Halifax *Herald* and Halifax *Evening Mail* started to write about what was occurring in Whitehead. As the seed of the co-operative movement started to grow, Capt. Billy Tom's articles and presentations appeared in co-operative magazines throughout Canada and were printed in the Co-operative League publications and periodicals in the United States.[12]

In 1938 the *Times* of New York sent a film crew to Whitehead to make one of their famous "March of Time" newsreels. In 1940 the newsreel was released internationally; however, it received little airing in Canada because of the continually breaking news of the events of the Second World War. Billy Tom's enthusiasm and matter-of-fact approach to the value of the co-operative movement, along with his candid insights into the struggles of the fishers, were being picked up by the local and national press and became a source of embarrassment to the federal and provincial agencies involved. It was not long before the provincial and federal governments, who had supported the formation of co-operatives yet were unwilling to provide funding, were forced to take a second look and get more engaged than they had been to date. It was not long before other ports followed suit and the spirit of co-operation among the various fishing communities that were hitherto isolated from each other began to flourish.[13]

CHAPTER 6

THE UNITED MARITIME FISHERMEN

When all is said and done, what is left when the fishers and farmers are taken away from these provinces? Their prosperity is synonymous with the prosperity of this part of Canada. If they are poor, we are poor. A few rich people will never make a civilization in these provinces. It is therefore the duty of all of us to move heaven and earth to better the financial conditions of the masses of our people, if we are patriotic citizens as we pretend to be.

– REV. DR. MOSES COADY, FROM HIS ADDRESS AT THE FIRST CONVENTION HELD TO INVESTIGATE THE FORMATION OF A FISHERMEN'S FEDERATION, JANUARY 1930

IN SEPTEMBER 1929 MOSES COADY SPOKE TO A NUMBER OF PEOPLE IN Canso. By the time he'd arrived, the fishers had already heard of the lobster sale made in Whitehead and the subsequent formation of their co-operative; however, they were still coming to grips with how they could solve their problems. They were unsure how the local fish merchants would react if they were to form a co-operative. Speaking later to a gathering at StFX, Coady explained that the Canso fishers weren't looking for handouts—what they wanted most was a viable plan of action and some support to implement it.

Coady set out on his tour of the Maritime fishing villages in the bitter winter of 1929–30. He worked extremely hard in the most difficult of circumstances attempting to educate fishers of the merits of the co-operative system and the need for study groups. For the most part, Coady's visits were well received. The fishers in the small Acadian communities along the Northumberland Strait, northern Cape Breton, Prince Edward Island, and the South Shore of Nova Scotia were more than willing to follow the lead of the fishers in Whitehead. As expected, Coady did have some pushback and reluctance to form co-operatives from the fishers along the southern Bay of Fundy in Nova Scotia and around the coast to about Clarks Harbour on Cape Sable Island. The major drawback was again the threat of reprisals from the local fish dealers if in fact their venture into the co-operative movement failed. The major dealers in the ports such as Lunenburg and Yarmouth were adamantly opposed to the formation of co-operatives and were quite vocal in their opposition. They made it very clear that they would neither support nor buy fish from any fisher that supported the co-operative movement. Yet in spite of these setbacks, Coady's notion was gaining momentum, and by the spring of 1930, he had gained the confidence of the Maritime inshore fishers to the point that he was prepared to form a fishers' federation.

Coady was convinced that by applying the principles of co-operation and education to the inshore fishery, this sector of the Nova Scotia fishery would recover and prosper. He knew that if the co-operative movement was to be successful, it needed a single over-arching focal point for the co-ordination of activities and a marketing arm for the various co-operatives that were now being set up in small fishing villages throughout the Maritimes. Coady also recognized the need for another focal point that would act as a voice of Maritime fishers and protect the inshore fishers and their families from any sort of retribution from the larger local fish dealers. With these goals in mind and from the input received from his meetings and discussions throughout the Maritime provinces, Coady drafted a constitution that would in time lead to the creation of an organization that would become known as United Maritime Fishermen (UMF). The question that remained was whether the individual co-operatives could come together to form a single organization.

Coady was well aware of the individual and independent nature of the fishers, but given their dire financial and social circumstances, he believed they would collectively come together within their own fishing communities or geographic area. The bigger challenge would be having the individual community co-operatives come together collectively under a single overarching co-operative venture. An additional concern for Coady was that for the co-operative movement to successfully consolidate under a single organization, it was imperative that it had the support of the federal and provincial governments and those businessmen and academics that were stakeholders in the Nova Scotia fishery. Historically, support from both levels of government was less than stellar, and those representing the larger private fish processing facilities had always been opposed to any changes to the status quo, especially if there was a new organization that would now compete with them in the marketplace.

In January 1930, Coady convinced both the federal and provincial governments to jointly sponsor a convention of sorts to discuss in detail the results of the recent Royal Commission and to open the floor to a general discussion on the state of the Maritime fishery. The convention would be held in Halifax and would be attended by a number of senior officials from the federal and provincial governments as well as academics and major stakeholders representing all various fishery enterprises across the three Maritime provinces. Mr. J. J. Cowie, the federal supervisor of fisheries, presided over the meeting. He introduced Coady to the assembly and paid tribute to the excellent work he was conducting in educating and organizing Maritime fishers. In his address, Coady reiterated the unanimous recommendations of the recent Royal Commission, foremost among them being that the fishers of the Maritime provinces become organized in line with modern business practices that were starting to prove successful in the manufacturing sector. He stated it was time for Nova Scotians, particularly those accustomed to the old economic order of things, to move forward if they wished to become competitive in the growing international marketplace. With respect to the shore fishery, he categorically stated that its very survival depended upon their ability to organize and speak with one voice. He acknowledged that not every facet of the fishery would agree with his analogy and recommendations, but he was adamant that this must be the way forward.

During his presentation, Coady provided the assembly with a number of examples from both Canada and abroad where the formation of co-operatives and the doctrine of co-operation made a significant difference and made a weak industry very strong. In his view, through co-operation and speaking with a single voice, tens of thousands of farmers and fishers in the Maritime portion of this country could be immediately put on their feet financially. Being able to become more financially stable and productive would in turn enable them to become buyers of all sorts of commodities used by other industries in the same region. It was at this point in the discussion that Coady set forth his doctrine and belief that the creation of the local organizations of fishers bound together by a single Maritime federation would provide a number of far-reaching benefits that would lead to more prosperity and a better way of life for fishers. He emphatically stated that if the shore fishery was to survive and move forward, then a Maritime federation must be formed that would allow an agency approach for the education of the people, act as an instrument for fact-finding in the development of general policies, and, in due course, take the lead in the development of a single co-operative marketing agency that could better exploit and consolidate market requirements. It was not surprising that Coady's address was well received and was given unanimous endorsement by federal and provincial authorities present and most of those representing the Maritime fishery. As expected, the only dissension regarding the organization of the shore fishers came from those representing the Grand Banks schooner fishery.

Having gained support for the formation of a single Maritime fishery federation from the federal and provincial governments, Coady's next step was to take this notion to the fishers themselves. In the spring of 1930, Coady, with the assistance of the parish priests and other local clergy, started to put together a meeting of representatives from all the newly formed co-operatives. Not knowing what the response to this venture might be, Coady forged ahead and sent out invitations to all concerned to attend what he called the "First Annual Convention of the United Maritime Fishermen." The meeting was to be held in Halifax on June 25, 1930, at the Masonic Hall.

To Coady's surprise and delight, over two hundred delegates representing all of the Maritime provinces and every aspect of the shore and near-to-shore fishery were in attendance. In addition, representatives from the freshwater fishery were also in attendance.[14] The Maritime shore fishery was about to go through the biggest and most significant transformation in its history. The decisions that were about to be made would affect it profoundly for decades to come and would lay one of the critical building blocks for the future development of the Cape Island–type longliner. Given the opposition that the large deep-sea fishing fleet owners and processors had toward the formation of the co-operatives, it was not surprising that there was no representation at the convention from the offshore banks fishery.[15]

In Coady's opening remarks he continued to stress the importance of the co-operative movement and the gains that could be made by working together. He stressed the need for a single marketing arm that could promote, market, and sell the catches of the fishers as well as become a single entity to voice the concerns of the fishers to the various levels of government. Coady then presented and tabled the draft constitution and explained in great detail the clauses of the constitution and how they would be applied to the new organization. In an effort to ensure that the constitution applied equally to the diverse fisheries of the individual provinces, each clause was reviewed in detail and amended if required. It was not surprising that those present unanimously approved the constitution. The organizational structure was to have the co-operative fishery divided into twenty-two different zones with a director from each elected to serve on the central board. There was agreement that the central headquarters, so to speak, would be located in Moncton, NB. An executive was chosen and the UMF became an official body.

Despite the enthusiasm for a central union, in the first year of its existence some critical issues surfaced that threatened to break the organization apart. The first president of the UMF, Chester McCarthy from Tignish, Prince Edward Island, always favoured a separate union and convinced the fishermen's co-operatives in that province to break away and form their own union. The Prince Edward Island fishers agreed they would co-operate with the main body on matters of co-operative structure and marketing but insisted that they would control their own policy matters. Although this arrangement met with some fear and skepticism in the beginning, it actually worked reasonably well, especially in the area of lobster processing and marketing.

It is one thing to sell an idea to people, but is quite another thing to be able to implement it to the point of success. Coady's skill and ability to organize the fishers into co-operatives had to be followed by a program that would provide the fishers in the co-operatives with the facilities to process, store, and transport their catch to the marketplace. While the federal government, in keeping with the recommendations of the MacLean Commission, did provide financial grants to StFX to assist with educational programs for fishers, up to this point, there had been no federal or provincial government financial capital to facilitate the construction of the much-needed infrastructure and more modern and up-to-date fishing craft. Coady knew only too well that if the required infrastructure, such as small fish-handling plants and wharves, was to become a reality, the fishers themselves would have to undertake the construction. It is not surprising that this hardy lot, who themselves were scratching for their very existence, turned—with their own limited resources—to the construction of small fishing stages and outbuildings needed to support their co-operative efforts.

It was believed that if a central agency could buy fishing supplies in bulk and distribute them, then the fishers could see considerable savings. Such a move would have the added advantage of the fishers being able to capitalize on the dividends distributed to the co-operative members from the annual profits being made on the sale of fish and supplies.[16] With no seed money from the federal and provincial governments, the individual co-operatives allocated some of their meagre profits to the UMF to facilitate some bulk wholesale purchases of netting, twine, hooks, etc. The fishers began to reap the benefits of this venture, as they were able to obtain their fishing supplies at a far better price than they ever did from the local fish dealers and chandlers. By 1939, the sale of fishing supplies to the individual co-operatives averaged around $12,000 annually.[17] There is no question that the local fish dealers and larger companies that had a monopoly over the price of fish and selling of fishing gear to the shore fishers were seeing a large decline in sales. This is not to say that the UMF was bent on putting the smaller dealers out of business; however, it did provide a strong measure of competition that brought the local dealers back to a point of reality and levelled the playing field for coastal fishers, especially in those communities that did not have or had not yet formed co-operatives.

In 1934, the UMF made the critical decision to go into marketing. In keeping with the principles of co-operation, it was believed that by consolidating the marketing effort into one central area, the individual co-operatives and the fishers themselves would receive a better financial return for the product. In the beginning the marketing division of the UMF had a hard hill to climb. In the very competitive international marketplace, the UMF was virtually unknown. However, as the co-operative effort expanded, especially with respect to the lobster industry in the Yarmouth area of Nova Scotia and northeastern New Brunswick, the UMF was gaining some good visibility for their products in the United States market. By 1939, the trade commissioners of a number of European and Caribbean nations were accepting the UMF as a viable producer with a good product and good financial stability. By the late 1940s the UMF had firmly established itself in the United States market to the point where 75 percent of all UMF catch (all species) was being sold through James Hook and Co., Boston, Massachusetts, and this remained so for a number of decades.

The UMF advanced its presence further by buying fish and other sea products from any fisher regardless of whether or not the fisher was a UMF member. This proved quite successful in those small communities where the population could not support a co-operative venture. One of the areas that the UMF was able to exploit was the annual catch of gaspereau, a relative of the herring that travelled up the Maritime inland waterways each spring to spawn. In many instances farmers, especially in Nova Scotia, harvested such runs and sold them to the UMF. The UMF in turn had a market in the Caribbean for a large quantity of the fish caught and had a local market for the use of gaspereau as lobster bait. Both the farmers and the UMF benefited, and the profits were in turn returned to the co-operative fishers. It is also interesting to note that some whose primary profession was farming became members of the UMF as fishers of gaspereau, shad, and smelt.

A good example of an early success of the co-operative movement and the UMF occurred in Alder Point, Cape Breton. Like most small communities, in the mid-1930s Alder Point was going through some troubling times. Most of the inhabitants were poor and barely hanging on. With the help and support of the local priest in the neighbouring community of Big Bras d'Or, the fishers studied the co-operative movement and were able to come together and subjectively examine their current situation and develop options. As a result of these study groups, the one hundred fishers in the area came together and formed the Alder Point Co-Operative Packers. In 1930 and without any government assistance, they undertook the construction of a two-storey lobster packing facility and fishmeal plant at a cost of $7,000. Some fishers put whatever savings they had into the project, and collectively they built and equipped this facility in hopes of eventually making a better life. The facility became fully operational in 1935, and with the help of the UMF central marketing organization, the fishers started to receive a good price for their lobster and, in the off-season, groundfish. By 1936, the plant was employing upwards of fifty-five young men and women. The financial impact this had on the community was significant. In that same year, the UMF reported that five additional canneries were opened, bringing the total number of facilities controlled by the association to twenty-two.

For the next forty-plus years, the original plant at Alder Point and its replacement prospered. The capital realized from the cooperative effort in the lobster market allowed the plant to expand into the processing of groundfish, and in the late 1950s onwards it was supported by a number of Cape Island–type longliners. Similar initiatives were undertaken in small fishing ports like Whitehead, Torbay, Chéticamp, and Port Bickerton to name a few, and although each facility varied in size and capacity, they became very successful in bringing a large measure of employment and prosperity to their respective communities and surrounding areas. The Port Bickerton operation commenced in 1944 with the guidance and support of Father Charles Forrest. In the coming years this plant prospered and grew to become one of the largest processing operations on the Eastern Shore of Nova Scotia. At its height in the mid-1960s, the plant was supported by twelve to fifteen Cape Island–type longliners, two medium-sized draggers, and a number of smaller fishing craft.

From the political perspective, the UMF became very involved in addressing issues of concern to the shore fishers. Each year the UMF held an annual convention and brought forward a number of resolutions that were in turn forwarded to the respective levels of government. Foremost among these was their continued affirmation urging a complete abolition of steam/motor trawlers operating out of Canadian ports. By the early 1960s this crusade started to fade as it became evident to some of the UMF plants that to remain competitive, trawlers would be required, and a small number were purchased by some of the larger facilities. However, the UMF remained very active in seeking the compulsory inspection of fish, better control of existing markets, and the need for sales agents in countries such as the West Indies to further market Nova Scotia fish.

From its founding in 1934 until its demise in 1989, the marketing arm of the UMF served its members well and was a force that balanced the needs of the shore fishery with that of the large corporations. Unfortunately, in the mid-1970s the problem of overfishing and the inability of the UMF to sell its product to an already flooded market contributed significantly to the eventual demise of its marketing organization. By the late 1970s the UMF was experiencing financial difficulties, and a number of the co-operative fish plants slowly started to close or be sold. The loss of the UMF caused a slow but steady erosion of the small individual co-operatives and the independence of coastal fishers.

CHAPTER 7

YOU CANNOT GET A BETTER FISHING CRAFT WITHOUT MONEY

A very large part of the fishermen of Nova Scotia have for years been endeavouring to earn a living with inadequate and obsolete equipment, a circumstance which not only reduced their potential catch while fishing, but also limited the variety of fish caught, scope of their operations, and the number of days they could work.

– MR. STEWART BATES, *REPORT OF THE ROYAL COMMISSION ON PROVINCIAL DEVELOPMENT AND REHABILITATION*, 1944

AS FAR BACK AS 1932, THE SHORE FISHERS IDENTIFIED THE REQUIREMENTS for larger and better fishing craft that would replace the small open skiffs and sail craft they possessed. The larger steam and motorized trawlers were taking their toll on the traditional inshore fishing stocks, and in spite of the restrictions placed by the federal government on Canadian trawlers, nothing prevented foreign vessels from fishing on the fringes of the coastal three-mile limit. With each passing day the shore fishery was getting harder. By the mid-1930s, progress was being made but the recovery was extremely slow.

The efforts of Coady and others were just getting underway when the Great Depression hit. This catastrophic economic event, although very difficult to overcome, did not stop the recovery effort, but it definitely had an impact on the speed of recovery. The formation of the UMF, as one fisher (my maternal grandfather) put it, provided a "bright light at the end of a long and dark tunnel," but both levels of government continued to stall and pay little attention to the immediate financial need of the individual shore fisher.

In the coastal fishing communities of Nova Scotia there was an increasing uneasiness among the shore fishers. It was a consensus that in order to successfully continue their recovery, there was a need for more able fishing craft and the ability to adopt better fishing methods, such as longlining. The fishers knew they could recover their livelihood and possibly prosper if given the opportunity to acquire better vessels and equipment. They also recognized that the shore fishery needed change, and as early as 1936 a number of fishers along the Eastern and South Shores of Nova Scotia, without any collaboration, independently started to identify what type of fishing craft they would require in the future and what form this vessel would take. In ten years hence, this vessel would become the Cape Island–type longliner. It must be understood that the development of a new type of fishing craft did not readily appear; as will be seen, it was an evolutionary process, and given the range of species being fished, there was never any intent to have such a vessel replace the existing fleet of fishing craft but rather complement it.

However, there were a number of practical issues in the mid-1920s to mid-1930s that prevented the development and construction of such a craft. The first issue was the vessel design. At this time, the vessels engaged in the shore fishery were very small coastal skiffs, the majority of which were sailing craft. The mix of fishing craft varied with the geographic area of the fishing communities relative to the species of interest and location of the coastal fishing grounds. A review of the vessels registered for the ports of Nova Scotia during this time shows that most vessels were between twenty and twenty-eight feet in length. Some were even smaller. (Note: In keeping with tradition and reflective of the time period covered by the book, imperial measurements, not metric, are used for boat dimensions.) By the mid-1930s, a significant number were constructed locally.

The completion of a lobster boat built by the author's grandfather at a local boatyard in Port Morien, c. early 1950s.
(AUTHOR'S COLLECTION)

Fishers could operate on the grounds close to shore, but by the end
of the 1920s the effects of the steam trawlers were taking a toll. By the
early 1930s, larger motorized craft were starting to appear in greater
numbers, but most still remained small and open to the weather and
incapable of venturing too far from land. Some were still based on
the schooner-type hull design, but the Cape Island–design fishing
craft, or vessels based on this design, were slowly gaining in popularity. By the late 1930s there was in some communities a very slow
but continuing trend to larger craft in the range of thirty to thirty-
six feet, while other communities, particularly along the Eastern and
South Shores of Nova Scotia, were going to the larger forty-foot vessel. At the time the number of larger motorized vessels remained

relatively low compared to the overall vessels in the inshore fleet; however, the number was increasing. A similar trend was occurring along the Northumberland Strait, but their vessel design, although similar in looks to the Cape Island craft, was somewhat unique in the actual construction and outward appearance due to the different sea conditions that existed in the area and surrounding waters. With the passage of time this design became known by the fishers as the "Northumberland build."

Fishers realized that to further advance their financial gains they required a vessel that could remain on the grounds for longer periods than just the day fishing, fish longer into the fall and possibly the winter, and be affordable. Longlining, which had been in use on the West Coast since 1913, was slowly being introduced on the Atlantic Coast and in the late 1920s and early 1930s was a relatively new method for harvesting groundfish in Nova Scotia. The knowledge base of the shore fishers with respect to fishing farther offshore, especially as it relates to the Atlantic fishery, had not matured. They knew what was required but lacked the expertise and ability to translate their requirements into a proper vessel design.

In the mid-1930s the formation of the UMF was finally allowing the shore fishers to make some financial gains, and progress although very slow was being made. With the help of the UMF, a good number of fishers were able to obtain motorized craft to replace the sail or oar powered skiffs. These vessels were small and limited in capability but were a quantum leap in "technology" when compared to what they previously had. But with the passage of time, the shore fishers were slowly gaining very valuable experience with longlining, and the requirements for the longliner were slowly being more clearly defined.

The second issue was the availability of suitable builders. Most of the medium-sized yards in Nova Scotia were still building vessels based upon the schooner design. It was not until the mid-1930s that the builders were slowly converting over to the construction of Cape Island–type fishing craft; however, you cannot build a boat for someone unless they have the money to pay. Shortly after the Great Depression a number of builders started to build small to medium-sized motorized fishing craft based on the Cape Island design, and by 1939 this design had evolved to what became known as the Cape Island "snapper" boat. As rudimentary as they may have seemed, there was a positive change starting to occur in the shore fishery.

A typical "snapper boat." (NS ARCHIVES)

The term "snapper boat" is a Nova Scotia fishers' term to describe a wooden Cape Island–type or similar fishing vessel that was normally thirty-eight to forty-five feet in length and engaged in the coastal and near-to-shore fishery. These fishing vessels were larger than the normal coastal lobster boats. The craft was usually of open construction and had the wheelhouse and covered space (or "cuddy," as it was called) located in the forward part of the vessel. The cuddy normally held a small accommodation space for a crew of three or four fishers. It was the forerunner to the Cape Island–type longliner. The term "snapper boat" was used to distinguish it from the local lobster boats that were normally smaller. The origin of this term, which is rarely used today, remains a mystery.

With the outbreak of the Second World War, the construction of fishing craft took a back seat to the needs for small to medium-sized boats for the navy. It is not surprising that the design of these support craft was founded on the Cape Island-type design. With Halifax, Sydney, and Louisbourg becoming major ports for naval activity, there was a need for small and medium-sized harbour craft, and the

Nova Scotia shipyards both big and small played a prominent role in addressing this requirement. As a result, the construction of new fishing craft over twenty-eight feet in length became limited, thus slowing the development of larger fishing craft such as the longliner and the overall recovery of the shore fishery. However, there is a positive note in that a number of Nova Scotia yards were involved with the building of larger wooden patrol boats that were totally different in design and structure from the traditional schooner. This provided valuable insight into the construction of larger conventional motor craft that would spill over into the construction of medium-sized motorized fishing vessels at the war's end.

The third issue ties directly into the first two and is perhaps the most important factor that precluded the early development of the longliner. This factor was the lack of financial support and resources available to the individual fisher. It was well known that to survive, the shore fishers desperately needed new and better fishing craft, better equipment, and a more honest and viable market. Unfortunately, the income and personal assets of the individual fisher at the time were insufficient to permit any type of loans or financial support from commercial lending institutions. This situation was further exacerbated by the lack of any financial support to the individual fisher from any level of government. It was not until after the establishment of the fishermen's co-operatives in the mid-1930s that fishers were finally able to get a proper price for their catch and started to see some financial gains. The need for change and better equipment was brought to the attention of and acknowledged by both levels of government as early as 1925, yet it took another twenty years, two Royal Commissions, and the aftermath of another world conflict to finally see some measurable progress toward the fishers being provided with much-needed financial support.

The establishment of small and medium-sized fish plants became a source of employment for those in the communities not otherwise directly involved in the fishery. However, a good number of fishers were still unable to obtain the necessary capital to purchase boats and equipment that would allow them to venture to the offshore fishing banks and fish later in the fall and early winter months. Both the findings of the MacLean Commission, released in 1928, and the results of the Dawson Commission, tabled in 1944, concluded that the shore fishery could neither grow nor flourish without some capital investment from both levels of government. The Dawson Commission was even more critical of all levels of government and categorically stated that if the Nova Scotia fishery were to survive, there had to be a process developed that would allow for loans and financial support to the shore fishery for new boats and equipment. Similar recommendations were made by the other five Royal Commissions that had, as part of their mandate, investigated some aspect of the Maritime fishery. However, the recommendations made by these commissions relative to the financial problems that were facing the fishery at the time were extremely broad in nature and as a result went unheard.

The events that unfolded in Canso on July 1, 1927, and that forced the Royal Commission of that same year did in fact strike a nerve with the local, provincial, and federal politicians. The media coverage that the co-operative movement was getting on the national stage was putting increasing pressure on the various levels of government to assist the shore fishers. The effects of the Great Depression of the early 1930s made any progress toward financial stability or providing financial aid to the Nova Scotia fishery extremely slow, but by 1935 the political pressure being exerted by the fishers forced the various levels of government to act.

Fredrick W. Wallace, editor of *The Canadian Fisherman*, was one of the strongest proponents for the individual fisher, yet in an editorial in 1937 he wrote, "If the fishing industry fails to attain a normal level of prosperity this year, it is not the fault of the federal and provincial governments." It would appear that he was supporting the notion that both levels of government were in fact providing for the needs of the shore fishers. Unfortunately, nothing could have been further from the truth. Wallace's statement provides a very good example of the federal and provincial governments' ability to politically manipulate financial initiatives to have one believe that they were indeed addressing the problems of the shore fishery when in fact the opposite was true. There was no question that in 1937 large sums of money were voted by both levels of government to aid the fishing industry and this money was welcomed. Despite the volume of funds expended, the portion allocated to individual fishers remained very small in comparison to the overall allocation. Although the Nova Scotia Fishermen's Loan Board was founded in 1936, it was not until the mid to late 1940s that it became an effective body with the ability and mandate to support the individual fisher. This body coupled with the full implementation of the Federal Fishing Vessel Construction Assistance Program in the early 1950s finally provided the long-awaited funds to the individual fisher.

Between 1935 and 1986, there were a host of federal and provincial financial initiatives that were very important to the expansion of the Nova Scotia fishery and that led to the development of the Cape Island–type longliner. That said, considerable financial support, although slow in coming to the shore fishery, was provided directly to the industry as a whole and was targeted toward the need for processing infrastructure and marketing, not to improving fishing vessels. Millions of dollars were spent supporting the construction of new wharves, freezer facilities, updated production lines and fish handling facilities, and the acquisition of new fish plants. There is no question that such facilities were desperately required; however, my review of all these initiatives revealed that throughout this period, the financial programs that would support the individual fisher in obtaining better vessels and equipment remained extremely limited. It was not until 1949 that any financial program of substance was made available to the individual fisher, and this program, the Federal Fishing Vessel Construction Assistance Program, had the single most direct impact on the recovery and expansion of the shore fishery and development of the Cape Island–type longliner.

Here are some of the government-sponsored and other programs that were available to the individual fisher from 1935 through to 1986.

NOVA SCOTIA FISHERMEN'S LOAN BOARD

In 1936, the federal minister of fisheries, Hon. J. E. Michaud, succeeded in getting a grant of $300,000 from the Dominion Treasury to be expended in the form of loans to needy fishers, provided the fisher's home province put up an amount equal to that granted by the Dominion.[18] Nova Scotia took $100,000 of that funding and matched it with a similar sum from its own treasury. The provincial Department of Trade, Industry, and Commerce was assigned responsibility for the overall administration of the program and established a specially appointed "board"[19] to oversee the approval of loan applications and disbursement of the funding. An amount of $150,000 was made

The term "bounty" refers to the payment made by the United States to Canadian fishers in 1873 as part of the Treaty of Washington signed between the United States and Great Britain. This treaty settled a number of longstanding disputes resulting from losses to American shipping caused by the British during the Civil War and losses to the British from the illegal fishing of American vessels in Canadian waters. With regard to the latter, it took some time to work out the details of a resolution. The British paid in excess of $15 million in reparations to the United States for the loss of ships and materials and granted the Americans permission to fish in Canadian waters. Sir John A. Macdonald was not in agreement with the decision and on his own negotiated a financial settlement for the use of Canadian fishing grounds. The Halifax Award, as the payment was called in Canada, was $4 million that was to be divided among qualifying fishers for the purchase of fishing gear or boats. What became known as the "bounty" was the annual interest on this money, to be divided among qualifying fishers. The regulations and structure of the bounty was spelled out by the federal government in the provisions of the Bounty Act of 1892. To qualify, a fisher had to show that they were engaged in the groundfishery for a specified number of days and had landed a specified amount of fish. In the day, it was a method of showing that an individual was a bona fide fisher. The bounty continued until 1966 when the terms of the original agreement were considered totally out of date and redundant. As a young man, the author qualified and received this bounty on two occasions.

available to "bounty" fishers (see sidebar) in seventeen Nova Scotia counties for the purchase of fishing gear and supplies for the hook-and-line fishery. Each county was allocated funding based upon the population of bounty fishers in that county. In each county, a local "loan board" was established to review and approve the loan application. Loans were not for the payment of back debts and were limited to $40 for individual fishers; in the case of two fishers or organizations, the loan was limited to $40 for each member. Fishers making application had to demonstrate that they were British subjects, had been a resident of the province for five years, were twenty-one years of age or older, and had three years of fishing experience. Another caveat to the loan was that the fisher could only obtain a loan in the county in which they resided.

Although the amount available to each individual was relatively small by today's standards, it was seen as a major step in improving the lot of the shore fishers. One has to remember that in 1936 the average annual income for most of the shore fishers was below $500, and a small motorized fishing skiff could be obtained for $40.[20] Also, for the first time, the loan was secured against the equipment purchased, not by the fisher's personal assets, such as his house and land.

Between 1936 and the release of the Dawson Report in 1944, the funding provided to this initiative grew. By 1942, funding appropriations had grown to where individual fishers were able to acquire loans amounting $450, but even with this increase, the amount of capital available to the individual fisher remained well below that needed for any major expansion or the purchase of vessels capable of fishing the middle ground.

Despite its shortcomings, the work of the Nova Scotia Fishermen's Loan Board was a start in addressing the financial needs of the individual fisher, and it had a positive impact on the shore fishery. Beginning in 1936 there was a steady annual increase in the number of more modern motorized vessels in the small ports along the coast of Nova Scotia. From the outset, the majority of vessels being built and financed under the loan board were smaller craft that had an average length of between twenty-five and thirty-six feet. The majority of these vessels were of open construction and were best suited for the lobster and the close-to-shore fishery. This was particularly true along the east coast of Cape Breton where lobster and handline fishing remained a lucrative but seasonal fishery. For those fishing ports from Canso to Yarmouth, there was a similar trend, but by 1941 larger vessels of up to forty-five feet started to appear in numbers as well. These vessels had the capability of engaging in the seasonal lobster fishery, rig for swordfish in the early to mid-summer, and continue into the fall and sometimes engage in a limited close-to-shore winter groundfishery.

NOVA SCOTIA FISHERMEN'S LOAN BOARD ACT

In its final report to the Nova Scotia provincial government in 1944, the Dawson Commission was extremely critical of the existing Fishermen's Loan Board for its lack of any formal legislative approval or oversight, its seemingly haphazard organizational structure, and the manner in which funds were allocated to the board and disbursed. The Nova Scotia government was forced to quickly address the concerns raised by the commission and in the same year, 1944, passed legislation that established the Nova Scotia Fishermen's Loan Board Act, which formalized the Nova Scotia Fishermen's Loan Board as a corporate body with its own funds. Originally, the government allocation to the board was $45,000. This amount was in addition to any federal grant monies that would have been matched by the province. In the years following its formation, the amount allocated to the board steadily increased, and by 1949 the amount available to assist individual fishers was in excess of $495,000.[21] Fishers began to take advantage of this funding, and with better boats and increased earnings, the risk of default steadily decreased. From the inshore fishers' perspective, the ability to increase their catch not only meant better personal prosperity but, given that a good number were members of the local fishermen's co-operative, the dividends received also allowed continuous improvement to their vessels and fishing gear.

VETERANS LAND ACT

The Veterans Land Act of 1942 was one of the rehabilitation measures planned by the Government of Canada to assist veterans upon the cessation of hostilities. From the outset, it was recognized that the returning service men and women would likely require some financial assistance to transition to civilian life. However, early on in the development of the program, it was recognized that the land settlement portion of the lands act would not meet the financial needs of those veterans who were seriously interested in full-time farming, commercial fishing, or smallholding settlements coupled with other employment. One of the principles of the benefits proposed in the legislation was to allow the veteran to apply for assistance to buy a home and re-enter the fishing industry. The benefit that was extended to those veterans engaged in commercial fishing fell under the provisions for property classified as a smallholding. A smallholding was defined as "seeking ownership of a property of one to three acres complete with a home that is suitable for this purpose on inspection by the administration of the loan and can be purchased from its present owner at a cost of $2,400." Once the purchase of the property was approved, the veteran could apply for advances of $1,000 to purchase a powered fishing boat and fishing gear.[22] To qualify, the veteran had to certify that prior to the war, he was engaged in the business of commercial fishing as a means of earning his living.

Under the terms of the act, the veteran could also apply for $150 for certain household items and equipment and $50 for small items and tools used in his business. The total funds disbursed by the federal government would be $3,600. For his part, the veteran was required to make an immediate payment of $240 or 10 percent of the cost of the land and improvements, whichever was the lesser. The veteran would then contract with the government to repay the sum of $1,600 over a period of twenty-five years at an interest rate of 3.5 percent per annum. The annual payment was set at $97, the equivalent of $8 a month. The capital cost absorbed by the government, or in other words the conditional grant to the veteran, was thus $1,760 or 48 percent of the total cost of the enterprise.

The act also provided a further type of assistance, namely funding to the veterans who already owned their home and desired funds to pay off the mortgage at a better rate, affect improvements to the property, or purchase equipment for their enterprise. Advances of funds totaling up to $3,200 could be provided for that purpose so long as it did not exceed 60 percent of the value of the land and other holdings. If the advance was for equipment only, then the total amount available would not exceed $2,500 or 50 percent of the value of the land and equipment. The land and holdings were held as security by the federal government, and repayment could be made over a period of twenty-five years at an interest rate of 3.5 percent amortized.

For the veterans leaving military service at the conclusion of the Second World War, this financial assistance was most welcomed. The majority of veterans were very young men, and after years of war they were looking forward to re-establishing themselves in the fishing industry and their local communities. A number of the veterans took advantage of this benefit and were able to build small homes and upgrade or purchase better vessels and equipment. Some took advantage of this benefit to support the acquisition of longliners in the early to mid-1950s. The provisions of this act remained in effect until 1965.

THE FEDERAL FISHING VESSEL CONSTRUCTION ASSISTANCE PROGRAM

In 1938 the Royal Canadian Navy clearly understood that if war were to break out in Europe, the majority of the naval assets based at the Canadian naval base at Esquimalt, British Columbia, would be transferred to the Atlantic theatre. It was recognized that the removal of such assets would significantly diminish the navy's ability to sustain a reasonable capability to patrol and defend the approaches to the Pacific coast. In an effort to counteract the removal of ships to the Atlantic coast, the federal government formed the Royal Canadian Navy Fishermen's Reserve. Under this initiative, a number of fishing vessels and their owners would be brought into this reserve force. Shortly after the establishment of the Fishermen's Reserve, the Royal Canadian Navy took possession of a number of fishing vessels, most of which were of the much larger packer/seiner type, enlisted a number of local fishing captains and crews, and, with very little training provided, had these vessels commence patrolling the vast Pacific coastline.[23]

The outbreak of war in 1939 and the need to provide food and supplies to the United Kingdom increased the demand for Canadian fish products from both coasts. The federal government believed that transferring fishing vessels to the Navy had created a shortage of fishing vessels on the West Coast at a time when fish products were desperately required in support of the war effort. This caused the federal government, under the War Measures Act, to provide a subsidy to West Coast fishers for the construction of new vessels.

On April 10, 1942, an order-in-council numbered P.C. 2798 was passed into law and provided for assistance to the West Coast fishery for the construction of packer/seiner-type vessels. The aid from the Federal Fishing Vessel Construction Assistance Program would be in the form of subsidy to be paid at a rate of $165 per gross ton for pack-ers and seiners constructed between seventy-two and seventy-eight feet in overall length and ninety to one hundred gross tons. The plans for such vessels had to meet the approval of the federal Department of Transport, and construction had to have been started before March 15, 1942. In addition to the subsidy, special provisions were made to the Income Tax Act for a special depreciation allowance of 20 percent per annum to the original owner for a period of five years.

In August 1942, the subsidy legislation was amended to allow for the subsidy to be applied to the Atlantic fishery for the construction of dragger-type fishing vessels and/or the conversion of existing schooners to draggers.[24] The rationale used by the federal minister of fisheries, Mr. Michaud, to support dragger construction was that fishers enlistments and the move of fishers to other war-related employment had sharply reduced the productivity of the Atlantic fishing industry at a very critical time. To be eligible for the subsidy, vessels had to be not less than seventy-two feet in length, and the work of construction/conversion had to have begun prior to July 1, 1942. What is interesting here is that this program on the East Coast was being directed to the larger producers and deep-sea fleets, not the shore fishery.

On June 1, 1944, the legislation was amended to include grants (at $165 per ton of the vessel's gross tonnage) to groups of fishers of not less than four for the construction of fishing vessels of the dragger or longliner type of not less than fifty-five feet overall length and not more than a maximum to be determined by the minister. Special depreciation allowances remained in force.

In that same year, the Royal Commission on Provincial Development and Rehabilitation was extremely critical of the failure of the federal government to recognize the importance of the fishery to the growth and stability of Nova Scotia. In light of the increased demand for fish by the European market, the government's failure to provide the capital investment necessary to sustain and modern-ize the industry, especially the shore fishery, was, in the view of the commission, inexcusable. It was equally critical of the seemingly pref-erential treatment that the West Coast fishery and the large fishing companies were receiving by way of subsidies for the construction of new fishing vessels with the dragger/otter trawler being the "pre-ferred" platform. The subsidy legislation of the time still did not allow individual ownership and as such placed the Nova Scotia shore fishery at a decided disadvantage. In his final report, Mr. Dawson strongly recommended that if the East Coast fishery was to survive and prosper, the subsidy program be continued at the end of hostilities[25] and be made available to the shore fishery. Similar recommendations were made by West Coast interests.

Under an amendment tabled in the House of Commons in 1947, Bill C-1919, the Federal Fishing Vessel Construction Assistance Program was retained and placed under the full control of the federal Department of Fisheries. At this time, regulations were amended to allow the subsidy funding to be paid directly to the provincial Fishermen's Loan Board in an amount not to exceed $165 per gross tonnage for the construction of fishing vessels of the dragger or longliner type. The subsidy also applied to schooners that could be converted to longlining. The only caveat in this particular amendment was that the subsidy was confined to vessels owned by a group of four or more fishers and measuring not less than sixty feet to a maximum to be determined by the minister. In accordance with the terms of this amendment, by definition a group of four or more fishers included co-operatives and incorporated companies. In the case of the latter, not less than 51 percent of the shares in such a company were to be purchased and held by not less than four fishers that had to be members of the crew of the vessel.

What is significant about this particular amendment was that although it now "formally" included the East Coast, it did not include any opportunity for the individual fisher to access the financial assistance. The amendment did, however, start to define what would constitute the requirements for a dragger and, of course, the longliner. In the case of the longliner, the plans for the vessel had to be approved by the Dominion Board of Steamship Inspection, and the vessel had to use a powered gurdy or trawl hauler during fisheries operations for each of the five years from the date of the issue of the inspection certificate. A review of the requirements set forth by the Board of Steamship Inspection further defines the construction and safety requirements for what was known as the "longliner class" of fishing vessel. In the succeeding years, this class of vessel became known by local fishers as the "government approved" longliner, which was separate from other vessels that also engaged in longlining.

In 1949, political pressure continued to be exerted by the Province of Nova Scotia and the province's media to reduce the vessel size required to qualify for the subsidy and allow individual ownership. In late 1950, the federal government conceded and approved the required amendment; however, the legislation was not formally approved until May 24, 1951. The order-in-council P.C. 2490/51 was issued that revised P.C. 1919/47 as follows: The reference to a "provincial loan board" was changed to "provincial government department or agency" authorized to provide assistance to fishermen by means of loans and otherwise.

It provided for the payments to the provincial department or agency of $165 per ton gross tonnage for vessels of the dragger and longliner type. These vessels could be owned by one or more fishers and measure not less than forty-five and not more than sixty feet overall length or be owned by any group of four fishers and measure not less than sixty feet to a maximum overall length to be determined by the minister.

The other conditions remained the same as legislated in P.C. 1919.

It was this order-in-council and amendment that not only defined the longliner from the federal standpoint but put in place a process and mechanism that would allow the common shore fisher to expand and go forward. Between May 1951 and the removal of the subsidies in mid-1986, there were a number of amendments to the subsidy requirements to suit the needs of the fishers and the fishery of the time. In 1958 an amendment was put forward that set the minimum length at forty-five feet and a maximum to be determined by the minister. This is somewhat significant as it allowed for the subsidy to be applied to the construction of a number of American-designed, sixty-five-foot longliners, seiners, and medium-sized wooden draggers with single ownership. From that date forward, additional amendments were made that relaxed some of the restrictions on vessel size, allowing for subsidies to be paid on smaller vessels such as lobster craft and also allowing some vessels up to one hundred gross tons be included in the subsidy program. What should be noted here is that the amendments of 1965 and beyond allowing subsidies on smaller craft had their own sets of construction and design criteria.

Although relatively late in coming, the amendments were welcomed especially by the shore fishers that were engaged in the lobster industry. Regardless of the nature of the amendments that were put forward during the life of the program, the definition of and standard for the construction of the "government approved" longliner remained as it was defined in 1947 and in the amendment of 1951. Vessels that did not meet these criteria, yet were granted federal subsidies under the various amendments since 1951, were classed by the federal government as being "multipurpose."[26] This financial program was the single most important contributor to the recovery and growth of the shore fishery and the birth of the Cape Island–type longliner. The program was cancelled in March 1986.

FISHING VESSEL INDEMNITY PLAN

In 1953 the federal Department of Fisheries authorized the Fishing Vessel Indemnity Plan to provide fishers with an affordable insurance program to cover the loss of or damage to their vessel. This plan came about as a result of significant losses to boats and equipment during the vicious winter storms of 1949. Prior to the approval of this plan, commercial insurance companies considered the risks associated with commercial fishing to be very high. As a result, commercial insurance was either not available or the cost of the premiums was out of the financial reach of shore fishers. With the implementation of this plan, insurance was made available to all coastal and freshwater fishers. Under the terms of the plan, insurance was available for vessels valued from $250 to $7,500. The premium for the insurance was 1 percent of the appraised value.

In 1957, the maximum value that could be insured was increased to $10,000 and increased again in 1958 to $25,000. As the program continued, it was evident that it had considerable merit. Fishing craft could now be indemnified up to 60 percent of the appraised value in the case a total loss, and up to 85 percent of repairs in excess of 15 percent of the appraised value. This plan worked well for both the federal government and the fishers. There were a number of amendments to the plan over the years, but as of 1985, the end date for my study, the plan was still in force.

FISHERIES IMPROVEMENT LOANS ACT

In December 1955 the Fisheries Improvement Loans Act came into force. The objective of this act was to increase credit to fishers and to facilitate the availability of intermediate and short-term credit for the improvement and/or development of fishing enterprises. When the act was originally established, the lending period was set at three years and was amended from time to time in later years to authorize additional lending periods. The term of the loan depended upon the amount borrowed, but by the mid-1980s a maximum term of fifteen years was established. The Fisheries Improvement Loans Act was originally administered by the federal Department of Revenue Bank and remained so until 1978 when it was moved under jurisdiction of the Department of Fisheries and Oceans. During the first few years of its operation, the number of fishers that actually took advantage of this funding was relatively small; however, participation increased, with the majority of those seeking funds under this act being owners and operators of medium-sized vessels such as longliners and draggers. The loans could be gained through any Canadian bank and were fully secured by the government. The government in turn held portions of the fisher's assets until the debt was fully paid. The funding made available under this act proved to be quite valuable to shore fishers in comparison to other forms of financing. The interest rate of prime plus 1 percent was rather attractive. The program was cancelled in June 1987.

PRIVATE SOURCES

The establishment of the UMF and the co-operative fish plants brought with it a limited source of capital in the form of small loans made by the UMF to its members. In some communities the co-operative movement expanded with the formation of Co-operative Credit Unions that provided another avenue for additional funding. The competition provided by the UMF and its member plants forced the private fish dealers in those communities that did not have a UMF presence to engage in similar activities with fairer and more competitive rates of interest, with the caveat that you sell your fish exclusively to that dealer until the loan was paid. What is interesting about this financial source is that beginning in the mid-1940s, it was most often used to help a fisher with the initial down payment to secure a loan for a larger boat from the Nova Scotia Fishermen's Loan Board. Ultimately, these loans would also be involved with the purchase of longliners and small draggers.

The financial arrangements worked something like this. A fisher would want to purchase a larger boat or even a longliner. In order to qualify for a loan from the Nova Scotia Fishermen's Loan Board, he had to provide 10 percent of the purchase price of the vessel. In the case of the longliner, in the early 1950s the price would be in the order of $20,000. This was a very large sum of money at the time, and it meant the fisher would have to provide approximately $2,000 up front to even be considered for the loan. If the fisher did not have the money, the local fish plant that would "buy" the engines for the fisher via a promissory note. The fisher would show that he had "equity" equivalent to the down payment and thus would qualify for the loan and the benefits of the Federal Fishing Vessel Construction Assistance Program. The fisher would take possession of a newer and larger vessel and would pay the fish plant back via personal arrangements. The processor/plant was happy, as now there was a larger vessel operating at the facility that in turn would mean larger catches and a longer fishing season. This small investment would add considerably to the overall profitability of the operation.

Finally, shore fishers had the means to buy a boat!

CHAPTER 8

THE LONGLINER: POLITICS, DESIGN, AND STANDARDS

*An elephant is a mouse drawn
to government specifications.*

– SOURCE UNKNOWN

DESPITE THE SETBACKS OF THE GREAT DEPRESSION AND THE catastrophic events of the Second World War, progress was being made in the shore fishery and the fishers were realizing some financial gains that during the 1930s would have been thought impossible to achieve. Unfortunately, running parallel to this success was the increased steam trawler presence on the traditional grounds of the shore fishery. Even before the end of the war, there was consensus among a number of fishers and co-operative fish plants that there would be a need for a medium-sized fishing craft that would complement the existing inshore craft and be capable of operating farther from shore and fishing longer during the year.

The introduction of the Federal Fishing Vessel Construction Assistance Program in 1942 became one of the two major cornerstones for the recovery of the shore fishery and the development of the Cape Island–type longliner. The second can be summarized with three words: politics, design, and standards. By 1947, the two cornerstones had become entangled and set off a firestorm of political

The FV *Helen C.* was a forty-foot coastal fishing craft built in Port Bickerton, NS, in 1937. This vessel is a good example of how other Nova Scotia builders were adopting the elements of the Cape Island design. (AUTHOR'S COLLECTION)

wrangling that to this day affects the relationship between Nova Scotia inshore fishers and the provincial and federal governments.

For the past two centuries, Nova Scotia has been world renowned for its shipbuilding. Starting in the late 1800s and continuing to the present day, thousands of fishing craft of various sizes, designs, and rigs have been built for and successfully engaged in the fishing industry along the Nova Scotia coast and the adjacent fishing banks. Some of the vessels were small, open, multipurpose craft, some were designed after the schooner-type hulls, some were converted schooners, and beginning in the early 1930s with the introduction of motorized propulsion, a great number followed the lines of the Cape Island design that had its beginnings in the early part of the century. Within the Cape Island design, a very special type of fishing craft started to evolve—a vessel that was born out of a need identified by the fishers themselves—and this vessel was the Cape Island "snapper boat." Its design ranged in length from thirty-five to forty-five feet, and yes, it was engaged in longlining for groundfish.

From this version of the Cape Island fishing craft came another more robust and larger variant that would become legendary before being almost forgotten. Of all the wooden Cape Island–type fishing

The class of a vessel is specific to its employment, whereas the type is specific to its construction and design relative to that employment. As an example, there are many different types or configurations of stern draggers, yet they all belong to the stern dragger class of fishing vessel. The same holds true for the longliner. However, with the longliner, the line between class and type gets quite cloudy when the actual "type" of fishing vessel is engaged in fisheries other than longlining. It was the federal government that, in defining the specifications for the various types of longliners as part of the Federal Fishing Vessel Construction Assistance Program, established the Cape Island–type longliner as belonging to a specific class. There was no singular design, and each builder developed their own "version" of the longliner; thus there were various "types," but each had to satisfy the requirements of the class.

vessels built in Nova Scotia for the shore longline fishery between 1949 and 1985, as far as I've been able to ascertain, only 205 can claim to be formally recognized by both the federal and provincial governments as belonging to a documented class of fishing vessel known as the "Cape Island–type longliner." To the fishers that owned and the crews that worked these vessels they became more commonly known as the "government approved longliners," as their design was formally approved by the federal Department of Fisheries as qualifying for financial support under the Federal Fishing Vessel Construction Assistance Program. Oddly enough this classification remains in effect today, but with the passage of time, the phase-out of wooden fishing vessels, the introduction of new construction materials, and different standards relative to the newer construction methods, the original intent of this classification has become lost and forgotten.

Establishing the Cape Island–type longliner as distinct from other vessels that engaged in longlining involves a review not of architects' or boat builders' plans but of the politics and intergovernmental wrangling that accompanied the implementation of the Federal Fishing Vessel Construction Assistance Program on the East Coast in 1942. Although the first Cape Island–type longliner was launched in 1947, a number of present-day fishers and others still argue about what constitutes the true definition of a Cape Island–type longliner. It is a little-known fact that there is indeed a formally approved definition for this type of fishing vessel, and this definition is not based solely on the vessel's employment but on a rigid set of design and structural standards that were developed and implemented by the federal Department of Transport in 1943 and reinforced by the federal Department of Fisheries when it assumed ownership of the Federal Fishing Vessel Construction Assistance Program at the end of the Second World War. These standards set the Cape Island–type longliner apart from its cousin, the large Cape Island snapper boat. To further define this class of vessel, and at the bidding of the federal Department of Fisheries, in 1949 the Nova Scotia Department of Trade, Industry, and Commerce developed and implemented a set of specific design standards that basically mirrored the federal standards. The only difference was that the standards were specific to the Cape Island–type longliner. As it turns out, the classification of the longliner and the rigid design and construction standards relative to the class are politically tied directly to the Federal Fishing Vessel Construction Assistance Program.

Beginning in the early 1940s, the federal Department of Transport had in place formal design and construction standards that defined the minimum specifications for the construction of fully decked wooden fishing vessels over fifteen gross tons. The standards provide very detailed engineering and structural data for fishing craft of specific lengths and rig. The standards are further broken down to ensure specific engineering and design criteria were applied to the various classes of wooden fishing vessels such as longliners and seiners.

In 1944, as a direct result of the Dawson Report and its scathing observations on the organization and conduct of the Nova Scotia Fishermen's Loan Board, the Nova Scotia Department of Trade, Industry, and Commerce issued a design specification and standard for the construction of open wooden Cape Island fishing craft ranging in length from thirty to forty-five feet. This specification was very general in its approach to the construction and was more or less developed to standardize the building materials (and their dimensions) to be used relative to craft of specific sizes. A review of the standards documents revealed that no specific reference or guidance was given to actual vessel structure, propulsion requirements, electrical systems, and crew comfort and safety, nor is there a requirement for any specific builder's trials. In fairness, most of the vessels built under this specification were registered as having a gross tonnage less than fifteen tons, and the majority were of open construction, that is, they were not decked aft of the wheelhouse at the time of construction and inspection. In essence, the intent of the specification was to provide some form of singular guidance for the revamped Fishermen's Loan Board to ensure that the vessels being built under the terms of the loan board were being constructed to a common standard. The federal construction standards that would be applied to the Cape Island–type longliner were far more detailed, with rigid structural and safety requirements imbedded. At this point, the Cape Island longliner had yet to be designed; however, the federal construction standards and definitions applied to the vessel's structure, not its employment. Therefore, any fully decked vessel fifteen gross tons and over had to be built to these standards.

The 1947 decision to continue the Federal Fishing Vessel Construction Assistance Program added a new dimension to the interpretation of these standards. In that year, legislation was passed by Parliament that transferred this program from the War Measures Act to the Appropriations Act and passed full responsibility for the administration and oversight of the program to the federal Department of Fisheries. One of the first steps taken by the department in implementing its control over the program was to add clauses to the qualification criteria that required a formal definition as to what constituted a dragger/trawler, a troller, a longliner, and others and established specific classes for each. A direct result of this initiative was that the provincial Fishermen's Loan Board was forced to satisfy the federal Department of Transport design documentation and the approval process that applied to the type of vessel being funded as well as the criteria established by the federal Department of Fisheries in order for the fishers to receive federal subsidy funding.

This issue was complicated even further when in 1947 the Department of Fisheries required each loan board to have their own design specification for each vessel type and class of fishing vessel being funded, and this specification had to match or exceed the design and construction standards administered by the Department of Transport and the Dominion Board of Steamship Inspection. In the case of new construction, if the vessel was to be granted the federal construction subsidy, the design documentation had to further define the specific class of fishing vessel to which it applied and specify what part of the fishing industry the vessel would engage in, for example, longlining or swordfishing. There was supposed to be no crossing the line. By regulation from the federal Department of Fisheries, the type of employment would define the vessel. To quote the words of one well-known builder, "this had the making of one hell of a row, and it did." To understand the complexities of this bureaucratic nightmare and the impact they had on the development of the Cape Island–type longliner and shore fishery, it is essential that one steps back in the history of Nova Scotia.

Since the beginning of the nineteenth century and continuing until the introduction of steel vessels, the quality of the Nova Scotia–built wooden vessels became world renowned. Just prior to and following Confederation, Nova Scotia–built vessels could be found in any port worldwide. A review of seafaring history will show that some of the largest sailing vessels in the world were built in Nova Scotia ports along the Bay of Fundy. The majority of the designers and builders of the time had learned their vocation in their home countries of Great Britain, Germany, and France and passed their knowledge on to succeeding generations, who also became world renowned builders. Even today some of their existing drawings in the various marine and national archives of Canada and those of other nations serve as a tribute to and examples of the craftsmanship that existed in Nova Scotia.

Despite their beauty, strength, and versatility, these ships were never subject to any formal governmental construction standards, inspection, or approval process. Instead, the vessels were built to a series of conventions that became a "standard" of sorts for vessel and small boat construction. For years master builders in Nova Scotia used and improved upon these conventions and methods, turning out vessels that were the envy of the world. The conventions were passed on to future generations of builders, and this practice continued until the mid-1940s. An excellent example of these conventions can be found in the book *Wooden Ship Construction* by noted naval architect and marine engineer W. H. Curtis, first published in 1919 and reprinted in 2009. Besides containing a wealth of construction details and methods, the publication highlights an issue that would raise its head some twenty years after its initial publication—the acknowledgment that every builder had a different approach to and interpretation of the conventions and methods. Put bluntly, there was no common standard for the design and construction of the vessels of the period.

A Nova Scotia–built Tern schooner docked at Port Greville, NS.

The larger ships such as the barques and Tern schooners were subject to a loosely administered approval process governed by British authorities, but this process was more related to the insurance of the cargo and crew safety rather than a standardized approach to construction and the ship's physical safety. With regard to the smaller vessels, especially those involved in the fishing industry, there was no standard nor is there any evidence that inspections were ever carried out by any governmental authority during construction.[27]

In 1868, a year after Confederation, the responsibility for all matters related to marine safety, ship surveys and registration, certification of shipping masters, etc., was transferred from the individual provinces to the new Government of Canada. In that year, the federal government formed a new ministry, the Ministry of Marine and Fisheries. Within this office, a directorate known as the Dominion Steamship Inspection Board was formed.[28] Although one of the prime functions of this directorate was to implement and oversee regulations related to ship safety, initially very little attention if any was paid to the actual structural design and construction methods used by builders of the time. Inspections of vessels under construction or repair were carried out, but for the most part they concentrated on the dimensional data for registration and ensuring that the vessel had adequate crew accommodations and safety apparatus such as navigational lights, anchors, dories/lifeboats, etc.

By the middle of the 1930s, the federal Department of Transport began paying serious attention to ships' safety as it pertained to the actual practices being used in ship construction. By 1940 Canada had developed a set of standards and inspection criteria for wooden and steel vessels. These standards were very stringent and were the first attempt to standardize inspection requirements nationwide. However, within the shipping industry there were no universal standards that would have international recognition. With respect to wooden vessels, there were a number of standards implemented in the US prior to 1943 that governed their construction and classifications; however, the standards were for the most part applicable only to the continental United States and the American ports bordering the Great Lakes.[29] Even within these standards there were a multitude of differences that caused variation and conflict among the coastal ports of the United States proper. Canada had developed a number of standards relative to the design, construction, and inspection of wooden vessels over fifteen tons, and not unlike the United States, in Canada there were major differences in the construction criteria among the individual shipyards in the Atlantic provinces, on the West Coast, and on the Great Lakes.

A similar situation existed in the United Kingdom; however, with the war still raging in Europe, the idea of imposing standards, although important, was all but forgotten. Notwithstanding the fact that the concentration of the Allied nations was on the war effort, there was a recognized need to define the limitation of the vessels' design based on sound engineering criteria, operational environment, and employment. A number of losses of wooden and steel vessels and their valuable war cargo as a result of poor construction prompted Canada, Great Britain, and the United States to attempt to develop specific design and inspection standards that would have international acceptance and recognition, but it wasn't until 1943 that any formal attempts were made. In that year, as a result of discussions between Canada and the United States, the American Bureau of

Shipping (ABS) was asked to host a major international meeting that laid the groundwork for the development of a comprehensive and highly detailed set of international marine engineering standards that would form the basis for a common set of design and construction criteria for vessels of all types constructed or repaired in the nations that were signatory to the adoption of the agreed upon standards.[30] By the time the meeting was arranged, the number of countries involved had increased significantly to include representatives from other European allies and South America.

The ABS was founded in 1862 in the United States as a non-profit society. Originally it was chartered to certify ships' captains, but it grew into an international body that became actively involved in the development and improvement of vessel construction and safety standards. Given its international reach, its ability to engage in every facet of the North American shipbuilding industry, and its lesser involvement in the ongoing war effort, it became an obvious choice to arrange and host the meetings in 1943.

Given the scope of the task, it is not surprising that within the ABS structure for the meetings, a number of subcommittees made up of noted naval architects, marine engineers, and master mariners from the member countries were formed to concentrate on specific areas of ship construction. Each subcommittee was further divided into various subgroupings that addressed specific aspects of vessel construction by purpose and type. With respect to the construction of wooden vessels, the main product that came out of the deliberations was a publication named *Rules for the Classification of Wooden Ships*.[31] Although the final document was based on a similar ABS document published in 1921, there were some major additions to address the advances in technology and concerns of the day regarding ships' safety. As a signatory member of the subcommittee, Canada, whose standards were more stringent than most, adopted a number of the ABS standards approved at the meeting as guidance for the development of better and internationally recognized wooden ship construction regulations.[32]

By early 1945 Canada had developed and implemented a set of regulations and standards that would govern the construction and inspection of wooden vessels between fifteen and one hundred tons; one hundred to two hundred tons; etc. The regulations and standards also created subdivisions that further classified vessels by their individual purpose and employment, and in some instances additional standards were developed to address the unique differences the type of employment created. In the case of fishing vessels, the tonnage requirements were also linked to the registered length, a practice that remains in force to the present day.[33] The resulting Canadian standards were laid down in a set of federal documents known as the Minimum Specifications for Building Wooden Fishing Vessels. In keeping with the agreements reached at the 1943 ABS meetings, all vessels constructed from 1908 until the implementation of the new regulations were exempt from the regulations and inspection requirements. All construction criteria used at the time of construction were to be considered compliant unless the vessel was subject to any major structural repairs or modifications. It was this set of regulations and standards that were to guide the development and construction standards of the Cape Island–type longliner and separate it from the familiar multipurpose vessels that hitherto had been involved in longlining.

A review of the Canadian regulations and standards produced in the mid-1940s revealed that for the most part the construction methods, the size and type of materials used, and the quality of fittings varied very little from what was considered normal practice for a vessel of a specific size and type constructed in a Nova Scotia yard. In fact, the construction methods being used in Nova Scotia had a major influence on the development of the ABS and Canadian standards. With respect to wooden fishing vessels, the new regulations laid out specific structural and safety equipment requirements that would be applicable to the type and purpose of the vessel under construction. The federal authorities viewed the smaller open vessels below fifteen tons to be limited in their ability to operate on the open ocean in excess of twenty nautical miles from land. Therefore, the regulations did not impact the small coastal lobster boats or the small to medium-sized Cape Island–type fishing craft. This is why a number of open forty-five-foot vessels constructed between 1945 and the introduction of the longliner were not subject to the same federal guidelines.[34]

When the Nova Scotia Fishermen's Loan Board was established in 1936, there was no formal construction or inspection standards by which to measure the quality of the vessel under construction. Most of the vessels constructed and funded under this program were below forty-five feet in length. The builders were well known to most of the

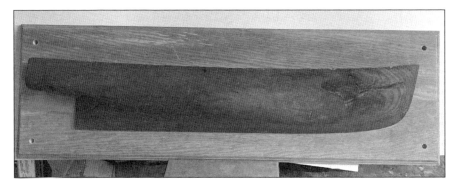

A typical builder's model. This model was developed by Clarence Heisler and was used to design and construct three Cape Island–type longliners under the Federal Fishing Vessel Construction Assistance Program. (AUTHOR'S COLLECTION)

members of the district or county boards of the time. The quality of the construction was consistently high and the approval of the board was based on an honour system of sorts.[35]

The majority of Nova Scotia builders designed their vessel from a scaled half-model. From this half-model, the builder would develop a set of working drawings that presented all the essential dimensional data needed for the construction of the full-sized vessel. The drawings, which also included a list of materials and hardware to be used during construction, were then enlarged to full size through a process known as mould lofting, and moulds were made for the shape and other parts of the vessel. When a fisher applied for a loan for the construction of a fishing vessel by a specified builder, the drawing would be forwarded as part of the application to the Nova Scotia Fishermen's Loan Board. Up until the middle of 1944, the engineering staff of the Nova Scotia Department of Trade, Industry, and Commerce deemed the builder's basic drawings and lists of material to contain sufficient detail to satisfy design criteria for the approval process. However, in that same year, with implementation of the Nova Scotia Fishermen's Loan Board Act, a formal set of design standards were developed for the construction of fishing vessels below fifteen gross tons. This initiative did not cause any difficulty as the standards basically mirrored what was being carried out by the builders of the time. Most builders would only have to submit a set of drawings for the first vessel.

Unless there were major changes to the design and profile of the vessel, the members of the Nova Scotia Loan Board would consider the builder an "approved builder" for the purpose of assessing the loan application; any change would be considered a modification of the original design. The implementation of formal design standards did not change this process very much as the standards were basically a formal document that captured what was being practiced by the builders as a normal course of construction. There is no evidence to show that the provincial standards were derived from any engineering study or ABS standards.

Prior to 1945, inspections were carried out on the small and medium-sized wooden vessels that had been financed through the Nova Scotia Fishermen's Loan Board by provincial authorities representing the applicable county loan board. The main focus of such inspections was to ensure that the vessel construction was in keeping with best practices, the vessel was considered seaworthy, and the terms of the contract were being satisfied.

In its original form, the Federal Fishing Vessel Construction Assistance Program implemented in April 1942 was very specific as to what type and size of vessel would be supported by the subsidy and as to the generous depreciation considerations for tax purposes. Given the need for resources to support Canada's war effort at home and overseas, and the need for seagoing resources to defend the West Coast, there was general agreement that those who had given up their vessels and/or volunteered their vessels and sometimes their crews to support the navy should have some form of compensation to allow them to return to their fishing enterprises at the end of hostilities. There was also a recognition that the navy's appropriation of vessels was having a negative impact on the West Coast fishery at a critical time when fish was required to support the war effort. At this point, Nova Scotia fishers saw the subsidy as another war-related initiative. Similarly, there was no question as to the need to provide some stimulus to the smaller British Columbia yards given the fact that a good number of the smaller yards in Quebec and the Maritime provinces were building smaller support craft for the Royal Canadian Navy and the marine service of the Royal Canadian Air Force.

However, in August 1942, just five months after the implementation of the Federal Fishing Vessel Construction Assistance Program, the attitude toward the program changed when the original legislation was amended to allow for the construction of draggers of not less than seventy-two feet overall length and the conversion of existing schooners to power draggers. In addition, the subsidy now "unofficially" applied to the Atlantic fishery.[36] This was considered a rather surprising move, as since 1929 there was a formally legislated set of restrictions on the use of steam trawlers on Canada's East Coast,[37] and there was a fear that allowing the buildup of a dragger fleet for use in the middle bank groundfishery would be just as bad as the steam trawlers. The argument provided by the federal government was that there was a need for more food products to support the war effort and the dragger was the most effective and required less personnel than other types of fishing craft.

In the eyes of the Maritime shore fishers and even those in central Canada, it appeared the subsidy legislation had gone beyond the bounds of supporting the war effort, and there was a perception that the federal government was once again giving preferential treatment to the West Coast fishery and the larger fish processors and producers. Skepticism grew when even more generous tax and depreciation allowances were given in the amendments of 1943. The shore fishers in Nova Scotia had a full appreciation for the need to support Canada's war effort but felt totally betrayed by both levels of government for their open support of the expansion of the steam trawler and medium-sized dragger fleet with no real support being considered for them. In 1942 and again in 1943, the United Maritime Fishermen unanimously passed a resolution calling for a complete ban on the use of draggers and steam trawlers from East Coast ports. Their concerns went unheard. Both levels of government and in particular some members of the provincial government of Nova Scotia were somewhat fixated on the need to expand the fishery by whatever means possible with preference being given to the medium- to large-sized motor and steam draggers/trawlers.

In early 1944 with the victory in Europe becoming more and more certain, the fishers of the Maritime provinces and Great Lakes were looking at how to rebuild their industries once the hostilities ended and take advantage of the new markets for fish products brought on by the devastation of the war in Europe. In Nova Scotia, both the shore fishers and the provincial government were outraged that support by way of loans or subsidies to the individual fisher was not even considered by the federal government, and once again the West Coast fisheries and the larger East Coast fish processing companies were seen as receiving preferential treatment.

In June 1944 the subsidy legislation was again amended. Under the terms of the amendment, the subsidy applied to the construction of fishing vessels of the dragger or longliner type of not less than fifty-five feet overall length, and access to the subsidy for these vessels was "formally" opened to the East Coast fishery. The fishers in Nova Scotia were pleased to see that the federal government appeared to be finally providing some sources of financial support; however, the subsidy remained available only to co-operatives, fish companies, and groups of fishers of not less than four. This did not in any way support the individual fisher who wished to purchase a small dragger-type vessel or longliner.[38]

The Nova Scotia Department of Trade, Industry, and Commerce shared the frustration of the fishers and made their concerns known to the federal Department of Fisheries. Unlike the central nature of the West Coast fishery, the Nova Scotia shore fishery was conducted from hundreds of small ports and was made up of a very large number of individual fishers who owned and mastered their own vessels. By the mid-1940s, and despite the national focus on the war effort, the co-operative movement now had a strong foothold in the province, and the growth of small co-operative fish plants was improving the financial lot of the small fishing communities. Even in those communities where co-operatives were not established, the private dealers had come to realize that the fishery had changed, and the competition among the co-operatives and other dealers had broken their long-held stranglehold on the lives of the shore fishers. If the private dealers were to survive, they too had to offer fair prices and support the need for better equipment and vessels.

To improve the growth of the fishery even further in such communities required that individual fishers have financial support to invest in larger and better-equipped vessels. The idea of having groups of fishers invest in a single vessel was unrealistic when one considered the demographics of the industry itself. The issue became even more mired in political controversy when provincial officials from the departments that were directly responsible for supporting the fishery were themselves split on the best way to use the federal subsidy. Some believed that the way of the future for the fishery should be a heavy investment in the dragger as it required less personnel while harvesting larger quantities of fish than could ever be produced by smaller longliners and the shore fishery. Others supported the need for a robust shore fishery and believed that the subsidy legislation as written would disadvantage the shore fishery and discourage individual ownership at a time when expansion of all facets of the fishery required the engagement of more and better-equipped vessels.

The Dawson Report of the Royal Commission on Provincial Development and Rehabilitation tabled in late 1944 perhaps settled this argument. In his final report, Mr. Dawson was very critical of the lack of support the federal government was giving to the Maritime fishery and in particular the shore fisher. He spared no words in stating that the subsidy program had gone beyond the bounds of its original mandate and was placing the East Coast fishery at a distinct disadvantage in comparison to the fishery being conducted on the West Coast. Mr. Dawson also stated that the East Coast fishery could never expand without the intervention and support of the federal and provincial governments and strongly recommended that the subsidy program be continued at war's end and include options for the individual fisher in the Maritime provinces, not just groups and companies.

In May 1945 the war in Europe came to an end. The devastation caused by this conflict and the urgent need to feed those poor souls caught up in it opened a large market for North American fish products. Canada recognized the need for a major expansion of the Canadian fishery, yet very little was being done to make this expansion possible. This lack of action was having a very negative effect on the ability of the Nova Scotia fishery to engage in and sustain any credible overseas markets. Meanwhile, the United States was expanding and becoming very aggressive in the marketplace at Canada's expense. In Nova Scotia the lack of support from the federal government prompted additional political pressure from the individual co-operatives and Reverends Tompkins and Coady.

The plight of the fishers did not go unnoticed by the press, and both the federal and provincial governments again came under fire for their seemingly "do-nothing attitude." The political pressure, especially on the federal government, to take some concrete action forced them to defer the cancellation of the Federal Fishing Vessel Construction

Assistance Program and conduct a review of the merits of continuance. This review was carried out and the results made public on May 22, 1947. The decision was made to continue the program.

From a vessel construction perspective, the construction and safety standards that were in force by this time would remain under the control of the federal Department of Transport, Dominion Board of Steamship Inspection. The subsidy remained open to fishers on the East and West Coasts as well as those that qualified in the Great Lakes fisheries. However, the subsidy was still only made available to vessels owned by any group of four or more fishers.[39] Although the basic principles and content of this new legislation remained similar to that of the program instituted in 1944, there were a number of additional requirements that were not included in the previous program. The policy of allowing a subsidy of $165 per ton for the construction of vessels of the dragger or longliner type remained the same, as did fishers' eligibility to access the subsidy to convert schooners to longliners. However, the funding was now applicable only to vessels measuring not less than sixty feet to a maximum length to be determined by the minister. The payment of the subsidy would no longer go to the owner/builder but to the provincial loan board and be applied directly to the overall purchase price of the vessel. Other major components of this legislation are as follows:

- For the first time, the program defined exactly what equipment each type of fishing vessel would carry to conduct its fishery. In the case of the dragger, the legislation specified the use of otter or other trawl of a similar nature, and in the case of the longliner, the vessel must use trawl lines and have a powered trawl hauler. The vessel could only be used for the type of fishery stated in the formal application.[40]
- Prior to the approval of any subsidy, a full and formal set of design drawings had to be approved by the Dominion Board of Steamship Inspection and the federal Department of Fisheries for either the dragger or longliner class of fishing vessel.[41]
- Prior to the release of any subsidy funding to the provincial loan board, the vessel must be constructed in accordance with the standards and regulations laid down by the Dominion Steamship Inspection Authority.

- The vessel was to be actively engaged in the designated fishing operations for five years from the date of issue of a Steamship Inspection Certificate by the Board of Steamship Inspection.[42]

A review of the legislation that governed the federal subsidy approval shows that this amendment makes clear reference to what would constitute a longliner, a dragger, etc. This requirement applied directly to the qualification criterion for the Federal Fishing Vessel Construction Assistance Program. The actual structural requirements of each were clearly defined in the federal construction standards; however, there are no references in the documentation to specify what type of vessel can be constructed, so long as a construction criterion is met. In essence, where an application for the federal subsidy was made, there were now two sets of requirements for each class of vessel. One set was based solely on the engineering and construction standards and would be administered by the federal Department of Transport, and the other, the financial assistance portion, would be administered by the Department of Fisheries. To qualify for assistance, both sets of requirements had to be met.

A major concern for the Nova Scotia Fishermen's Loan Board was the federal requirement for "approved plans." This requirement could be easily achieved on the West Coast where vessels from the various classes were readily identifiable and followed a similar design and rig, and where approved plans had been in place since 1943. However, on the East Coast the situation was totally different. The Nova Scotia fishery had many different designed craft that were engaged in longlining and other fishing activities.[43] Provincial officials as well as the shipbuilders were unhappy with the federal government's insistence that the current builders' drawings were inadequate and only drawings from certified naval architects would be acceptable for construction under the Federal Fishing Vessel Construction Assistance Program. When questioned on this matter, the federal government provided a typical political answer, stating that this requirement applied only to vessels that made application to the federal Department of Fisheries, and the loan boards were free to use such drawings for vessels in which they alone had financial interest.

Since the introduction of federal construction standards for fishing vessels in 1945,[44] the Dominion Board Steamship Inspection had been conducting compliance inspections on all vessels below sixty feet in length and over fifteen tons constructed and/or modified in Nova Scotia yards and had no difficulty in accepting the builders' drawings, list of materials, and construction methods. Between June 1944 and the 1947 amendment allowing singular ownership, only a small number of vessels were built and granted the subsidy; they were primarily built for the herring fishery and were classified as seiners. During these formative years of the inspection process, there is no documentation to indicate how thorough these inspections were. What is known is that after 1947 the inspections became more stringent.

Prior to the 1947 amendment, the federal authority responsible for administering the Federal Fishing Vessel Construction Assistance Program had no difficulty in allowing the Nova Scotia Department of Trade, Industry, and Commerce to use the builder's drawings to grant loans and approve builders for various types of fishing craft and grant the subsidy for those vessels that met the qualification criteria in place at that time. However, with the introduction of the latest amendment and the transfer of the subsidy program to the federal Department of Fisheries, requirements suddenly changed and it appeared this department wanted more control over the vessel approval process.[45] What justification did a federal ministry that was not responsible for enforcing safety standards have in insisting that the shipyards provide formally approved drawings? Why was there no objection from the Department of Transport? The Nova Scotia Department of Trade, Industry, and Commerce and the fishers saw this latter move as being political to the core. The immediate impact of this new requirement was a disqualification at the federal level of a number of well-known and reputable building yards from participating in the subsidy program until their drawings were formally approved by a "federal engineering authority." Even more important was that the disqualified builders were the only ones that had the knowledge and capability to construct the vessel so needed by the shore fishers. A good example of this occurred in January 1951 when the fishing vessel *Elizabeth & Jennie* was launched from the Clarence R. Heisler boatyard at Youngs Island (now Gifford Island), NS. This vessel met the structural and material requirements for the longliner class of fishing vessel as defined by the Dominion Board of Steamship Inspection. It also met and, in some cases, exceeded the provincial standard for the Cape Island–type longliner. This vessel, which in all respects could be called a Cape Island–type longliner, was approved by the Nova Scotia Fishermen's Loan Board based on the builder's drawings yet was not eligible for federal subsidy assistance because the drawings were not formally drawn by a "certified" naval architect.

The number of naval architects available in Nova Scotia at the time was very limited, and the cost of the drawings would have to be borne by the builder.[46] The issue of drawings was further complicated when those responsible for the subsidy at the federal level insisted that before they would grant a subsidy to any prospective client, the loan board itself must approve drawings and standards for every type of vessel that would be subject to subsidy approval and that this approval would have to be measured against a provincial construction standard for the class of vessel involved. This requirement added more confusion to the construction process and was interpreted by the Nova Scotia government to mean that any wooden and decked fishing vessel with a minimum length of sixty feet and over fifteen tons built in Nova Scotia would require the approval of the Dominion Board of Steamship Inspection prior to being eligible even for provincial loan funding. This argument was countered by the federal government stating that it had no jurisdiction over any loans granted by the government of a province; however, the vessel would not be classified by them as being of the longliner or dragger class, would still require compliance inspection, and would not qualify for the federal construction subsidy. This meant, therefore, that the Nova Scotia Loan Board could still function and approve loans for dragger, longliner, or seiner-type fishing craft, but before the federal government would grant the subsidy for construction of the vessels, design documents would have to be developed for each type and submitted to the federal government for approval.

At the time of the 1947 amendment, there were a number of approved drawings and standards available for the medium-sized coastal draggers sixty feet and over that in some instances were used for herring seining. The approval of the design of these vessels and the associated drawings was not as problematic as the majority were based upon small to medium-sized inshore draggers that emulated American and New Brunswick designs and had been signed off on by recognized naval architects. The problem that faced the Nova Scotia government was the development of a design standard for a new type of vessel that would satisfy the needs of the inshore fishers engaged in longlining in the near-to-shore and middle-bank groundfishery.

CHAPTER 9

ALL AROUND THE MULBERRY BUSH

The 50 foot Unique, *which can be built for a cost of about $20,000.00, was designed as a working model for a projected fleet that may well decide the destiny of inshore operations in the Maritime provinces. It was a joint project of the Nova Scotia Department of Trade and Industry and the Federal Department of Fisheries.*

– THE CANADIAN FISHERMAN, MAY 1960

ONE OF THE PROBLEMS THAT SLOWED THE RECOVERY PROCESS FOR THE shore fishery was the fact that both levels of government treated the Nova Scotia fishery as a single entity and did not separate the shore fishery from the offshore. Therefore, when it came to the gathering of data there was no recognized separation between the shore and offshore fisheries, with the end result being that the large production numbers of the offshore fishery masked the conditions that existed in the shore fishery. It is therefore not surprising that from a government perspective the fishery was doing well. The same issue existed when it came to examining new fishing methods and fishing craft.

Despite its growing popularity, versatility, and potential, the Cape Island–type design was not initially given any serious consideration by the Nova Scotia Department of Trade, Industry, and Commerce as a fishing platform for the shore to middle bank fishery. Beginning in the late 1930s and continuing until 1949, the federal Department of Fisheries and the Nova Scotia Department of Trade, Industry, and Commerce believed that the answer to addressing the need of the shore fishery for a better fishing craft was to follow the lead of the West Coast and American fisheries by having a standardized or common basic design. There would be some variations depending upon the builder, but these could be considered minor modifications with no impact on the structural standard requirements. The thought process of the time was that having a common basic fishing platform would allow for a single set of approved design drawings that could be easily modified to accommodate future changes to fishing practices and equipment. A single dragger/seiner design (or a modification of it) was already being used in the herring and scallop fishery in the Bay of Fundy area fishery with reasonable success. Having a standard design for the longliner would compel each participating builder to stick to a standard configuration that from an administrative standpoint would simplify the approval process for the granting of provincial fisheries loans and facilitate subsequent inspections. With the introduction in the early 1940s of more stringent design and construction standards and the Federal Fishing Vessel Construction Assistance Program, the notion of a common design gained more traction with each level of government getting involved. Despite their willingness to finally help, the programs that were introduced had a major flaw: they failed to consult with the fishers.

This notion of a common fishing platform design resulted in a number of separate initiatives undertaken at the provincial and federal levels. In addition to examining fishing platforms, both levels of government independently conducted a number of experiments using various types of fishing gear and investigated other types of marketable fish outside the traditional species. In 1936 the federal government conducted some minor research in the areas of longlining and trawling, but it was not until 1944 that they built what they believed would be a new experimental longline fishing boat that had the potential of becoming the common design for use in both the shore and offshore fishery.[47]

The vessel was the MV *J. J. Cowie*. Built by the Industrial Shipping Company in Mahone Bay, NS, the *J. J. Cowie* was designed on a modified Pacific coast longliner plan.[48] It had an overall length of 65 feet, a beam of 15.3 feet, and a gross tonnage of 48.76 and was built with

MV *J. J. Cowie* after its major refit, ca. 1950. (FISHERIES COUNCIL OF CANADA)

the sole intent of testing the usefulness of this design on the Atlantic coast. In its original configuration, it had a wheel and deckhouse located forward, with a workspace abaft of the deckhouse, a design feature common to the Pacific fishery. There were some similarities with the Cape Island design except that the vessel had a very limited sheer, and the placement of the wheel and deckhouse came further aft.

Despite the optimism expressed by the federal government officials responsible for the *J. J. Cowie*'s construction, the fish plant owners, fishers, and even UMF executives openly criticized the construction of the government vessel. It was their contention that the forecastle was too long and the stern deck, which was the most important feature for working, was too short. Of equal concern was that the main hold was not large enough for good fish runs. In examining the vessel's lines, the fishers and others firmly believed they would experience considerable difficulty in conducting fishing operations from the *Cowie* in Atlantic waters. If this were to be the case, then a complete redesign would be required.

Unfortunately for the federal government, the criticism was not unfounded. The *Cowie*'s design proved to be less suited for the Atlantic fishery than originally anticipated, and its overall dimensions were too large for the small port shore fishery. In 1949–50, the

J. J. Cowie underwent a major refit and was extensively modified. Its whole upper deck and engine room space were completely re-designed. During this refit, its forward deckhouse was removed and a new wheelhouse was placed in the after part of the vessel. Its final deck design was similar to some of the dragger-type fishing craft that were being built in the United States.

Although the shore fishery never accepted the original design, the MV *J. J. Cowie* was one of the first government-sponsored experimental vessels on the East Coast. In spite of its original design flaws, after modification the *Cowie* became a very valuable asset as it undertook a number of experimental initiatives that benefited both the shore and the offshore fishery.

The Nova Scotia government was aware and most supportive of the federal initiative, but like the fishers and others, believed that a vessel the size of the *Cowie* might not be practical for the shore fishery. In all fairness to the Nova Scotia government, there was considerable merit in exploring a common design, as such a practice was already in place along the Eastern Seaboard of the United States. Most of the medium-sized vessels operating out of the New England ports had a single design and had the capability to either engage in longlining or be rigged for coastal and/or close-to-shore dragging. In fact, a number of vessels of this design were built in the late 1950s in Nova Scotia yards for the herring industry.

Oddly enough, the single design initiative continued to be pursued in Nova Scotia well into 1947 when the Department of Trade, Industry, and Commerce sponsored the design and construction of a multipurpose vessel christened the MV *Shirley Adeline*. The vessel was based on an American fishing craft originally designed by naval architects from Boston and had an overall length of forty-five feet. This American design was commonly used for the inshore fishery along the Eastern Seaboard of the US. The Nova Scotia Department of Trade, Industry, and Commerce bought the rights to the drawings from the builder in the US and had them modified to meet the federal standards imposed by the Dominion Board of Steamship Inspection. In modifying the design, provincial authorities followed the American practice of having the vessel capable of being configured to engage in longlining, dragging, or purse seining. The *Shirley Adeline* was built at the L. H. Hirtle Shipyard in Mahone Bay, NS, for Captain Peter Fiset, Chéticamp, NS.

A model of the FV *Shirley Adeline* which was made before the ship was built at the L. H. Hirtle Shipyard, Mahone Bay, NS. (NS ARCHIVES)

It was hoped that this vessel could be the one selected as a common design to meet the needs of the longliner/small dragger for the shore fishery. Unfortunately, like the *Cowie*, the *Shirley Adeline* did not perform in the waters off the coast of Nova Scotia as well as it did on the fishing banks off New England. The craft was originally rigged for close-to-shore dragging but was found to be totally underpowered for this task. It then was converted to longlining; however, the vessel was found to have poor seakeeping qualities, especially in its area of operation—the waters of the Gulf of St. Lawrence. After a few short years, Capt. Fiset sold the vessel in favour of a Cape Island–type longliner.[49]

The end result was that this design was no longer explored for use in the Atlantic shore fishery but did have some limited success in the Bay of Fundy area as a seiner. Again, in fairness to the provincial government, their design was reworked by a number of noted Nova Scotia builders with the result being a very successful medium seiner/scallop-type fishing craft.

In their quest for a single design, the provincial Department of Trade, Industry, and Commerce made three glaring errors:

- The imperative of the time and more so after 1947 was to find a vessel that would satisfy the needs of the shore to middle bank fishery and that would already have an approved set of structural drawings that would satisfy the requirements of the Federal Fishing Vessel Construction Assistance Program, especially those of the federal Department of Fisheries. Unfortunately, they pursued the single design initiative without any serious consultation or discussion with Nova Scotia shipbuilders and/or fishers, particularly the shore fishers.
- The province investigated vessel designs that were being built outside Nova Scotia and used along the Eastern Seaboard of the United States instead of looking inward to the Nova Scotia shipbuilding industry.
- Very little attention was paid to the gaining popularity of the Cape Island design.

There is perhaps some justification for this latter point, as during the mid-1930s when the notion of a common design was first broached, the majority of the Cape Island fishing craft starting to appear in Nova Scotia ports were forty feet and fewer in length. Most were still open to the weather and used as an open, multipurpose fishing craft in the lobster and near-to-shore fishery. It wasn't until the 1945 to 1949 time frame that the Cape Island vessel began to expand in length, width, and draft and take on the attributes that would finally define the Cape Island–type longliner. The governments of the Maritime provinces had no difficulty with the implementation of construction and safety standards that were administered by the federal Department of Transport; however, the question was then raised that if the federal Department of Fisheries had no specific design requirements that would identify each class of vessel (i.e., dragger, longliner, etc.) other than the equipment used for a specific fishery, why was there a need for the provincial loan boards to develop a specific design document that identified the individual vessel type for each category of employment? As a result, the Maritime provinces believed the most expeditious way of resolving the matter was to have a common, pre-approved design.

The political and bureaucratic wrangling over standards and design further increased tensions between the federal and provincial governments. Nova Scotia now faced the daunting task of developing a design standard for fishing craft that had to be a minimum of sixty feet in length and at the same time come up with a design proposal that would allow the approval of the schooner and dragger/trawler-type hulls for longlining prior to these vessels being granted a subsidy. The development of an actual design was perhaps the easiest part, as a number of yards were already building medium-sized Cape Island–type fishing craft, and to increase the length was not that difficult. The challenge would be to satisfy the requirements of the federal Department of Fisheries.

The shore fishers and some larger fish processors in Nova Scotia looked upon the 1947 amendment to the federal subsidy program as a backward step. Not only did the vessel minimum size requirement increase, but again no consideration was given to extending the subsidy to single owners, even though the Nova Scotia Fishermen's Loan Board had approved loans to individual fishers since its inception in 1936.[50] The fishers also believed that considerable time had been lost in chasing various designs that from the beginning were deemed unacceptable, and the solution was already staring them in the face: the Cape Island–style fishing boat. The need for design documentation seemed totally political in nature. In some communities, disappointment gave way to anger as the restrictions on the subsidy were seen to favour the larger companies and the construction of medium and large draggers.

Like the fishers, the Nova Scotia government believed that the subsidy was a major first step toward expanding the Maritime fishery and providing a means for the rapid recovery of the shore fishery; however, neither saw how this program in its current state would be of any real benefit to the individual fisher and the shore fishery as a whole.[51] The amendment had not only increased the vessel size requirement for subsidy qualification, but caused a change to any formal provincial standards that were under development.

In late 1947, the restrictions on the construction of trawlers/draggers were formally lifted. A number of the original drawings and designs that were developed by noted architects in the New England states were available and being used by Nova Scotia builders. To modify the drawing to accommodate longlining from these vessels was rather simple—the inclusion of a gurdy house aft of the forward accommodation space. However, a number of builders, especially the smaller ones that had developed their own designs, were disadvantaged. Like the construction of the Cape Island–type vessels of the time, half-models and builder's drawings were available and prior to 1944 were approved by the Nova Scotia Fishermen's Loan Board. But because these plans did not contain the structural data and were not certified by a naval architect, they were not acceptable for federal subsidy approval.

It is not surprising that despite the thrust to expand the Maritime fishery, very few draggers and dragger-type longliners were built in Nova Scotia between 1947 and 1951. Like the fishers of the time, the builders felt that they too were being played as political pawns in the disagreements between the provincial and federal governments. In the eyes of the governments of the three Maritime provinces, there was no question that the expansion of the Maritime fishery would have significant economic impact on their respective provinces, and the Federal Fishing Vessel Construction Assistance Program was seen as a major factor in allowing the expansion to happen. There was unanimous agreement that the criteria being imposed by the federal Department of Fisheries for subsidy qualification seemed to fly directly in the face of the reasons for continuing this program. This latter point set off a furor of discontent and animosity between the federal and Nova Scotia governments that would last for years.

In 1950 the increased political pressure from both the Maritime and West Coast fishers finally drove the federal government to amend the subsidy legislation to more adequately reflect the needs of the individual fisher. In that year, the federal Department of Fisheries introduced a revision to the subsidy legislation as follows:[52]

- The amendment provided for payments to the provincial department or agency of $165 dollars per ton for vessels of the dragger and longliner type.
- The vessel could be owned by one or more fishermen and measure not less than forty-five and not more than sixty feet overall length (see sidebar), or be owned by any group of four or more fishermen and measure not less than sixty feet to a maximum overall length to be determined by the minister. All the remaining conditions were to stay the same.

The **registered length** of a vessel is the length recorded in the vessel's official registry. It is the length of a perpendicular line from the fore side of the stem where it joins the keel to the after side of the head of the stern post or, in the case of a ship without a stern post, to the fore side of the rudder stock. The **overall length** of a vessel is the distance from the foremost part of the stem/bow to the farthest point aft at the stern. The overall length is normally greater than the registered length. Why the federal government used the overall length rather than the registered length as a standard remains a mystery.

Up until the 1950 amendment, only groups of fishers could apply for the subsidy (i.e. the vessel would have to be owned jointly) and the vessel minimum length was longer. When on May 24, 1951, the government passed an order-in-council that treated the revision as a formal amendment to the Legislation of 1947 and passed it into law, this amendment finally opened the door and paved the way for the individual shore fisher to have access to the subsidy program and consider the construction of small longliners and draggers. The small and medium-sized shipbuilders also welcomed the amendment, as the subsidy would definitely increase the demand for smaller vessels, thus increasing their business.

To the Nova Scotia Department of Trade, Industry, and Commerce, the change was welcomed news; however, there were still two major problems that had to be overcome. To accommodate the change in vessel size for subsidy qualification, it was necessary to have the design drawings for the dragger/schooner-type hulls approved by the federal Departments of Transport and Fisheries. This would require provincial intervention to assist those builders that had to have their own half-model designs properly translated into the required engineering drawings. Given that most of the hulls were based on a proven or approved design from drawings and specifications already in existence, the approval process could be easily accommodated by showing any different configuration as a "modification" of the original design so long as the technical and structural standards remained intact. The federal Department of Fisheries and the Dominion Board of Steamship Inspection accepted this proposal.

Somewhere between the release of the subsidy amendment in 1947 and 1950, the Nova Scotia Department of Trade, Industry, and Commerce started to recognize the rapidly increasing popularity that the Cape Island–style fishing vessel was gaining in the shore fishery and the potential the basic vessel design presented for developing a larger version that could be used by the shore fishery for working farther offshore. In 1949 the Department of Trade, Industry, and Commerce set to work to develop a construction and design standard for the Cape Island–type longliner. Unlike the dragger/schooner hull design, there was a design but no existing, formally approved design drawings for the Cape Island–type longliner that incorporated the structural requirements of the federal Department of Fisheries and Dominion Board of Steamship Inspection. There was no question that Nova Scotia builders were meeting the structural requirements, standards, and inspections of the federal Department of Transport. In fact there is evidence to show that some of the structural requirements adopted and laid down by the Department of Transport were developed and in use in Nova Scotia shipyards during the construction of schooners as early as the 1930s and during the major construction boom of support craft for the RCN and RCAF during the Second World War. However, the federal Department of Fisheries was not willing to approve subsidies for the Cape Island–type longliner without a provincial design document that clearly identified the vessel type and class.

The job of developing this standard became the responsibility of William H. Hines, chief engineer of the Fisheries Division, Nova Scotia Department of Trade, Industry, and Commerce. The resulting document, called the "Detailed Specifications for Construction of 55'6" Long Liner Type Fishing Vessel," mirrored the structural specification and design criteria laid out by the Dominion Board of Steamship Inspection for a vessel of this size and purpose. A review of the federal specifications and design documents and those produced by the Nova Scotia Department of Trade, Industry, and Commerce shows that the only difference between the two documents was that the nomenclature and safety requirements were applied directly to a Cape Island–type hull. It is indeed ironic that the provincial government was required to submit a set of specifications and design criteria for each class of vessel that had been implemented and exercised by the Dominion Board of Steamship Inspection since 1945.

It can be said with some degree of certainty that the only require-ment for the provincial specification was for the sole purpose of sat-isfying the federal Department of Fisheries.

With the development of this specification, the Nova Scotia Department of Trade, Industry, and Commerce had satisfied the requirements of the federal Department of Fisheries for the release of subsidy funds to approved applicants. The specification would be the guiding document for the construction of any forty-five to six-ty-foot wooden Cape Island–type longliner for the next thirty years. Over the years there have been a number of amendments to these construction standards, but most have been implemented to address environmental and safety concerns along with the introduction of new and better engineered building materials.

By the late 1950s there were some indications that the provincial government was pressing for a federal subsidy for smaller fishing craft. There was a recognized need for financial assistance for the new, smaller fishing craft coming into service between thirty and forty-five feet in length. Most were employed in the lobster fishery but capable of engaging in other fisheries on a smaller scale on the fishing grounds closer to the coast. Unlike the larger longliner, these vessels varied in design depending upon their geographical location and employment outside the lobster fishery.

In 1959 the notion of a common design for fishing craft was still alive in the eyes of the Government of Nova Scotia. In 1960 a new type of fishing vessel was launched at the Stright-MacKay Ltd. yard in Pictou, NS. The design of the vessel was to fill a perceived void between the thirty-foot and forty-five-foot inshore fishing boat and the longliner and dragger. It was designed by the National Research Council, was classified as being totally multipurpose and capable of engaging in any facet of the close-to-shore fishery, and was launched with considerable political fanfare. Unfortunately, the vessel was not readily accepted by fishers and builders and received considerable criticism from both. Given the trouble the government had had in satisfying the federal Department of Fisheries's requirements for the Cape Island–type longliner, when the notion of a single common design arose again, it had the same negative results.

Finally, in 1965 the federal Department of Transport took the step to incorporate the standards for the multipurpose class of the Cape Island design and the smaller Cape Island–type longliner into

FV *Unique* launched at Stright-MacKay Ltd., Pictou, NS, 1960.
(FISHERIES COUNCIL OF CANADA)

a single publication called the "Minimum Specifications for Building 35–50 Ft Wooden Fishing Vessels." With the passage of time and advancing technology, a number of products that were stated in the original specification were proven to be harmful to one's health and were removed from the specification and replaced by safer products. Similarly, with the introduction of better navigational aids, radio communication, firefighting equipment both fixed and portable, and lifesaving equipment, the specification has been amended to reflect these changes. What has not changed are the standards for the Cape Island–type longliner relative to the structure and the basic require-ments for strength and endurance.

It is also important to note that between 1951 and the closure of the subsidy program in 1986, there were a number of significant amendments made to the subsidy legislation to accommodate the needs of the day. The amount of subsidy paid to each individual or to the provincial authority changed significantly over time, as did the minimum and maximum lengths of the vessel subject to the subsidy. Yet through all these changes, no further reference or amendment was made to the original definition and the basic requirements for the wooden Cape Island–type longliner as a class of fishing vessel. Although antiquated by time, the definitions and the classification have remained in force to this day—thus the term "government approved longliner."

With all the federal requirements finally satisfied, the shore fishery was now able to acquire and operate the modern, medium-sized fishing craft that they had fought to have for a considerable time. The day of the Cape Island–type longliner had finally arrived.

CHAPTER 10

CAPE ISLAND-TYPE LONGLINER: BIRTH OR EVOLUTION?

I got to tell you, lads, in my experience in being around these 'longliners' in my days in Lunenburg, North Sydney, and Port aux Basques, all over the Atlantic Canada, you could not find a better fishing vessel in the longlining fishery. There are some close relatives of these vessels, but the build and seaworthiness, speed—a real workhorse of the era.

– CHESTER HOWARD H. LLEWELLYN, RETIRED FISHER

ON AN OVERCAST DAY IN MAY 1950, THE FV *DAVID PAULINE* SLID gracefully down the slipway at the Warren Robar Yard in Upper LaHave, NS, and entered the placid waters of the LaHave River. Such an event was commonplace to those who lived in the towns and villages where seacraft were built, but this event was different. The *David Pauline* was touted as being a new type of fishing vessel—a "government approved longliner," built purposely for the expanding inshore fishery. It was an astounding 55.5 feet in length, had a breadth of 15 feet and a depth of 4.5 feet. Its overall design was based on the Cape Island fishing vessel that had been steadily gaining popularity since it first appeared in Cape Sable Island in 1905.

Launch of the FV *David Pauline.* (FISHERIES COUNCIL OF CANADA)

The launching of the *David Pauline* was given considerable attention in the local media and soon gained national attention. It was reported that the design of this vessel resulted from the joint cooperation of the builder, the engineers at the Nova Scotia Department of Trade, Industry, and Commerce, and the Nova Scotia Fishermen's Loan Board through which the vessel was financed. It was built for Captain Lawrence Bolivar, Liverpool, NS. Mr. William Hines, a naval architect and chief engineer of the fisheries division of the Nova Scotia Department of Trade, Industry, and Commerce, headed the overall initiative. It was he who was given most of the credit for the vessel's final design.

In June 1950, *The Canadian Fisherman* magazine published a very detailed article on the design and features of the *David Pauline.* In addition to the vessel's physical dimensions, some of the features included in the design were that it was fully decked aft of the forward house, a fish handling/gurdy house was fitted separate from but just aft of the main wheelhouse, a mechanical trawl hauler was installed in the gurdy house, and the crew accommodations forward of the wheelhouse allowed for extended trips to the fishing grounds. In addition, two six-cylinder Acadia Marine gasoline engines using a twin-screw configuration powered the vessel.[53] Perhaps the most important feature of the craft was that its design was approved by the federal Department of Transport, Steamship Inspection Division and followed the specifications developed by the Nova Scotia Department of Trade, Industry, and Commerce for the class of fishing vessel known as the "Cape Island-type longliner." According to Mr. Hines, the vessel met the requirements of the federal Department of Fisheries for a financial subsidy under the Federal Fishing Vessel Construction Assistance Program, which paid for the construction of new fishing vessels on the basis of gross tonnage.[54]

Since that eventful day in 1950, the *David Pauline* became known as the first of a revolutionary type of Nova Scotia fishing vessel and the first "government approved" longliner built in Nova Scotia. Its very design was perceived to be the standard that would be the basis for all future longliner construction. In fact, every longliner from that date, regardless of design and/or builder, was referred to as having been based on the "Robar" design. Unfortunately, this has been a long-held misconception. First and foremost, every decked fishing vessel over fifteen tons was subject to the design criteria laid down by the Board of Steamship Inspection, and second, the term "government approved" was used in direct reference to meeting the requirements of the Federal Fishing Vessel Construction Assistance Program.

As we shall see, historical documentation reveals that the design of the Cape Island-type longliner did not magically appear from the drawing board of any one naval architect but was the result of an evolutionary process that used as a baseline the Cape Island-style fishing vessel. It was the fishers that identified the requirements, and it was they who established the essential "operational" criteria that in the end would define the Cape Island-type longliner. The requirements that were identified by the fishers coupled with the implementation in the mid-1940s of federal construction and structural requirements took the basic design of the forty-three-foot Cape Island-type longliner one step further.

To understand how the Cape Island-type longliner came to be, one must look beyond the politics that surrounded the Federal Fishing Vessel Construction Assistance Program and examine the evolutionary process that transformed a small but rugged powered fishing skiff into a fishing vessel that would become a Nova Scotia icon.

By the mid-1930s, the serious problems that plagued the shore fishery through the first two decades of the twentieth century were slowly but methodically being addressed. It is a historical fact that during this period in Nova Scotia's maritime history there was considerable political wrangling at all levels of government over issues affecting the fishery; however, it must be acknowledged that there was some positive recovery. The Great Depression of the early 1930s slowed progress, but despite the economic hardships of the time, the shore fishery was experiencing some modest growth. Funding was being provided via the co-operatives to allow some fishers to obtain better fishing craft, and the "inshore fleet," so to speak, was seeing some positive growth. The co-operative movement and the establishment of the UMF were slowly starting to impact the small coastal communities in a positive way, and small local fish plants were starting to appear in a number of ports along the coastline of Nova Scotia. The availability of a more structured and comprehensive marketing strategy coupled with the ability of the fishers to process catches locally at fair market value allowed them to expand their activities.

In the early 1930s, the lobster fishery was still the major source of income for most of the smaller fishing communities in Nova Scotia. Between 1900 and the late 1920s, this fishery was conducted very close to shore using either small sailing craft or hand-rowed skiffs. By the end of the 1920s, sail was giving way to a somewhat larger motorized vessel, and foremost among them was the Cape Island boat. The vessel started its evolution soon after its humble beginnings at Clark's Harbour in 1905. By the beginning of the 1930s, this design was becoming well known and was gaining in popularity throughout Nova Scotia, and a number of variants from the original design were being constructed. In some areas, designs similar to the Cape Island design were starting to appear, each having the hallmark of its builder. From the outset, most of these vessels were relatively small open craft with an average length of twenty-two to twenty-eight feet. Over time most were powered by a single- or dual-cylinder gas engine commonly known as the "one lunger." Although the introduction of these motorized craft marked a quantum leap forward, they were still nothing more than a powered skiff, open to the weather, with limited capacity and versatility.

The after-run of a hull refers to the shape of the vessel from its centreline to the stern. In the case of the conventional schooner, the actual hull was deep and rounded, and this shape followed through all the way to the stern. With the introduction of engines, the propulsion power was now being applied via a propeller located very close to the stern and ahead of the rudder. The power generated by the propeller caused the conventional hull shape to "squat," making the vessel somewhat difficult to handle while tending fishing gear, especially lobster traps. The Cape Island design, unlike the schooner, had a shallower draft and a wider and square stern. The shape of the hull below the waterline was flatter from about midships to the stern. This change prevented the stern from squatting with the application of power and made it easier to handle.

Coincidental with the introduction of the small marine engine came a growing interest in the automobile engine. The increased power and versatility of the automobile engine forced another significant change in the original Cape Island design. The physical size of the engine and gearbox forced the Cape Island boat to increase in length and breadth. The length of the vessels started to increase to an average of thirty to thirty-six feet, and breadths of eight to ten feet soon became the norm. The after-run and lines from amidships aft were made flatter and broader, and the stern was given a wider transom. The long flat run of the underwater hull prevented the vessel from squatting under the increased power provided by the internal combustion engines, while at the same time allowing more working space amidships for fishing gear.

The Cape Islander now possessed the capability of going a bit farther offshore, especially during the summer months, and allowed fishers to increase the amount of fishing gear. In the case of those involved with the groundfishery, more trawl lines equated to longer lines of gear, thus the term "longlining" started to appear in the local vocabularies. At some point during this phase of the vessel's evolution, some enterprising fishers cobbled together surplus automobile transmissions and differentials and connected the assembly to the main engine via a belt to fashion a mechanical hauler—the forerunner of the trawl gurdy.[55]

Cape Island–style fishing craft at Cape Sable Island, NS. Note the installation of the forward cuddy, windbreak, and in some cases the open wheelhouse. The photo also shows the installation of mechanical trap and trawl haulers. (NS ARCHIVES)

Other features started to define the Cape Islander even further. Some vessels were modified to have a small shelter called a "cud" or "cuddy" attached to the forward structure of the boat. These small forward cabins were originally designed for the shelter and secure storage of personal and small articles of fishing gear, but it also provided some shelter from the wind. Some fishers along the South Shore of Nova Scotia took advantage of the structure and added small windscreens to further increase the weather-keeping capabilities of their vessel. It was not long before the builders started to raise the forecastle deck to allow for a larger cuddy. A permanent wheelhouse replaced the windscreens on some of the larger vessels. By the beginning of the Second World War, the vessel had developed from its original design as a small, open, motorized fishing skiff to the thoroughly versatile fishing craft whose lines are still recognized worldwide. The ancestors of the Nova Scotia longliner were born.

Commencing in the early 1930s and continuing to 1947, there were three significant events that profoundly influenced the capabilities of the Cape Island boat and led to the design of Cape Island–type longliner. Ironically enough, the first was the shore fishers' expansion into the swordfish fishery in the mid-1930s. This expansion did not encompass all of the fishers, but the numbers and the economic impact that those engaged had on the total shore fishery was indeed significant. What is more startling is that there is a direct correlation between the fishers who participated in this fishery and those who would own and operate the Cape Island–type longliners fifteen years later.

The swordfish has been a transient species in the coastal waters off Nova Scotia for centuries. The Atlantic swordfish winters in the warm waters off the Carolinas and migrates northward to feed in the cooler waters of the western Atlantic during the summer months. In the early part of the twentieth century, spring brought with it an abundance of herring, mackerel, and in some locations capelin to the coastal fishing grounds and harbours along the eastern coastline of Nova Scotia. The presence of these species attracted great numbers of swordfish to our shores. The fish would start to appear off the coast of Yarmouth and surrounding fishing banks in mid-May and migrate northward through June and July as far as Cape North in Cape Breton, returning to southern waters in mid-September. In those days, the swordfish could be found very near to shore out to a distance of twelve to fifteen nautical miles, making it accessible to the shore fisher using small to medium-sized fishing craft.

The swordfish is not a schooling species but swims alone and during the summer months can be found basking near the surface of the water. Up until the beginning of the 1960s and the introduction of the swordfish trawl, the most common fishing method was harpooning the fish while it was basking. There was always a market for the elusive "broadbill," but to harpoon a fish from a small sailing vessel was extremely difficult and inefficient. With the introduction of the gasoline engine into the Cape Island–type fishing vessel, the fishers now had a relatively fast and manoeuvrable platform that was ideal for catching swordfish, and the expansion of the shore fisher into this market began. By the mid-1930s until its closure in 1971,

Swordfishing boats from ports along the eastern shore of Nova Scotia fishing out of Port Morien, NS, in the early 1940s. (AUTHOR'S COLLECTION)

the swordfish fishery was one of the most profitable enterprises for a host of small ports along the eastern coast of Nova Scotia and was one of the major "unofficial" contributors to the development of the Cape Island–type longliner.

The staggered lobster seasons along the Nova Scotia coast coincided with the migration of the swordfish. It was not uncommon for a fisher—for example from the Eastern Shore community of Port Bickerton, NS—to complete the local lobster season in late May to mid-June, take advantage of the mackerel that were abundant on the shore at the time, and then rig out for swordfishing in early July.[56] By the mid-1930s and on into the mid-1940s, a good number of fishers from ports like Cape Sable Island, Lockeport, Port Mouton, Port Bickerton, and Canso, to name a few, started to follow the migrating swordfish northward and spend their summers fishing out of ports in Cape Breton such as Louisbourg, Port Morien, Glace Bay, Dingwall, and Ingonish.[57] This migration or increased mobility of the swordfishers forced another significant change to the Cape Island design. The fisher had to create living quarters in the cuddy. The requirement to carry one or sometimes two twelve-to-fourteen-foot dories

Fishers from the local port and from ports all over Nova Scotia comparing notes in Port Morien, c. 1942. Everett Feltmate, the author's father, is seated, centre, with his sleeves rolled up. (AUTHOR'S COLLECTION)

necessitated a further lengthening of the hull as well an increase in breadth amidships and a wider stern. Vessels started to appear in increasing numbers with a length of forty to forty-five feet, a breadth of twelve to fourteen feet or more, and a much wider stern.

The second important influence was the migration of the fishers themselves. Commencing in the 1930s until the late 1970s and the introduction of fiberglass and composite materials, there were a large number of small and medium-sized shipbuilding yards located along the entire coastline of Nova Scotia. Although the majority of fishing vessels built were small, most followed the basic Cape Island type. Each builder added his unique touch to the basic design, and each builder's design was easily recognized by the fishers. As the vessels from along the southern and eastern shores of Nova Scotia started to intermix with similar vessels from Cape Breton, fishers and builders began to identify various features in the fishing craft that could be incorporated into future vessels. The end result of this intercourse of ideas was a number of variants that followed the basic Cape Island design, but each had an increase in strength, versatility, and efficiency.

The third and most important influence was the impact the Canadian and American draggers and seiners had on the inshore fishery. From the beginning of the late 1800s until 1970, Canada's territorial waters extended only three nautical miles from the Nova Scotia shoreline. In most ports along the mainland Nova Scotia, Cape Breton, and Bay of Fundy coastlines, the fishing grounds worked by the small fishing craft extended beyond the three-mile limit. In 1927, one of the recommendations made by the Royal Commission Investigating the Fisheries of the Maritime Provinces and the Magdalen Islands was to place restrictions on Canadian-owned motor/steam trawlers operating from Nova Scotia ports and close these ports to foreign trawlers. The basis for this recommendation was the recognized need to protect small inshore fishers and coastal fish stocks. The end result of this recommendation was the implementation of restrictions versus an all-out ban on the number of Canadian-owned steam trawlers and those operating from Canadian ports. Despite the Canadian effort to protect the shore fishing grounds from the larger Canadian trawlers, the American trawlers and seiner fleets continued to operate within three miles of the Canadian coastline.

Between 1930 and 1945, the impact American and some European trawlers/draggers operating on the grounds close to the Canadian mainland had on the stocks of groundfish and bait fish was being felt by the small-boat fishers up and down the coast of Nova Scotia and the Bay of Fundy. Although the shore fishery had started a slow but steady recovery, the declining stocks made this a very steep hill to climb. From 1943 onward, the decline in the mackerel, herring, and capelin stocks caused a steady decline in the number of swordfish and other groundfish species being caught in the near coastal waters off Nova Scotia. The November 1951 issue of *The Canadian Fisherman* reported, "Nova Scotia fishermen edged their craft back into home ports during early October to mark the close of the 1951 swordfishing season. The season's success remains a split decision where record kills were reported in offshore waters while inshore landings were the leanest in years." Swordfish and other species were now being found in offshore waters, with the larger stocks being in and around the outer fishing banks and Sable Island, some 180 nautical miles from the closest point of mainland Nova Scotia. The movement of the swordfish farther offshore meant that the small vessels with a length of twenty-eight to forty feet could no longer engage in this fishery. This was a significant financial blow to an industry that was just getting back on its feet. Similar trends were being experienced with the traditional cod, haddock, and halibut stocks. Handlining and trawling from small vessels along the shore were still profitable but not to the extent they had been a decade before. Like the swordfish, the large stocks were now on the middle banks and too distant to allow the traditional small open boats to engage. If the fishers were to travel to these banks, larger boats with a capability of staying at sea for longer periods and working in an extended fishing season would be required.

For the fishers, going to a larger fishing craft created its own dilemma. The majority of the Cape Island–type boats that had been engaged in the swordfishery for the previous decade were for the most part multipurpose vessels. These vessels could engage in the seasonal lobster fishery, tend the mackerel and herring nets, and allow some longlining to be conducted during the spring and early fall after the swordfishing season was over. Increasing the size of the vessel would make it too large and cumbersome to be used in the close-to-shore lobster and groundfish fisheries. The movement of the swordfish farther offshore meant that even the open forty-five-foot Cape Island boats were approaching the limits of safety and endurance.

The increased demand for fresh and salt fish brought on by the Second World War and the decade following caused the shore fishery itself to split. Beginning in 1945, the lobster catches along some portions of the Nova Scotia coast from Halifax to Chéticamp on Cape Breton Island started a period of decline. In some ports along the Eastern Shore of Nova Scotia, groundfish and swordfish slowly became the staple of their fishery. A number of the fishers, seeing an opportunity, broke away from the small-boat fishery and invested in larger vessels.[58] The requirement now was for a vessel large enough to be able to travel safely to the Sable Island area during the swordfishing season as well as being able to engage in the fall, early winter, and early spring longline groundfishery.

This requirement identified by a number of the fishers around the province caused a further change in the design of the traditional Cape Islander. Vessels started to be built with a length of forty-five to fifty-two feet. To accommodate bigger engines and fuel, the vessels were becoming deeper with a wider beam and stern. Watertight floors were becoming the norm for the workspaces aft of the wheelhouse. The freeboard amidships was also becoming higher. Although the sheer of the vessel changed very little, the wheelhouse was being extended further aft to allow for shelter when working trawl lines. In some instances, this shelter was being separated from the wheelhouse by a partition, thus forming the nucleus of what would become the separate gurdy house. The traditional accommodation was also increased to allow the vessel to stay at sea for three or more days. The open boats were now being decked with removable watertight hatches to allow for better watertight integrity aft of the wheelhouse and more hold capacity and to shelter the ice that was now carried to help preserve the catch during extended trips to the fishing grounds.[59] By 1947 some vessels were being launched from Nova Scotia shipyards fully decked aft and with a shelter aft of the pilothouse for handling gear.

Not all of these shelters were separated from the wheelhouse by a watertight bulkhead, a critical requirement for the longliner class of fishing vessel as defined by the Dominion Board of Steamship Inspection. A good number were coming off the ways with an open configuration only to be decked after launch. These vessels, however, received the proper certification for their class, which at the time was "multipurpose." Such vessels, which provided much more flexibility and endurance, became known to the fishers as the "snapper boat." The fishers that invested in this type of fishing craft were, depending upon their geographic location, able to engage in the spring lobster and groundfish fisheries, rig for swordfishing for the summer months, and then convert back to longlining or lobster fishing for the fall. The only major drawback was, depending upon the locality, the ability to engage in the winter fishery. A number of vessels did so on a limited basis, and it was quickly realized that the limits of this design had been reached.

By the mid-1940s, the shore fishers realized that the Cape Island–type vessel was ideal as a fishing platform, but to those that were looking to work farther out to sea, the craft had to be more robust, be fully decked aft of the wheelhouse, be watertight, have more power, and most importantly, be affordable. Commencing in the fall of 1948 and continuing into January 1950, three fishing vessels were constructed at the Clarence R. Heisler shipyard, Youngs Island, NS. The vessels were the FVs *Janet Louise*, *Zoomer*, and *Zilch*. Mr. Heisler designed these vessels from his own half-model, and all were based on the Cape Island-type hull. There is historical documentation that disproves the notion that the *David Pauline* was a revolutionary new type of fishing craft and proves that the Cape Island-type longliner came about through an evolutionary process that cumulated in the first of those three vessels designed by master builder Clarence R. Heisler.

Unfortunately, due to its size and the nature of the qualification requirements in place at that time by the federal Department of Fisheries, the *Janet Louise* could not qualify for financial assistance under the Federal Fishing Vessel Construction Assistance Program and therefore could not be classed as a "government approved" longliner. While it could be argued that because the vessel did not qualify for the subsidy it could not have been considered a longliner, nothing could be further from the truth. Documents reveal that the vessel indeed met all the critical standards as defined by the Dominion Board of Steamship Inspection for the longliner class of fishing craft. More importantly, these standards were directly applied to the Cape Island-style fishing craft. My research has revealed that at this time there were a number of similar Cape Island-type fishing craft built, but this was the first time that a Cape Island-type fishing craft had met or exceeded all the federal standards for the longliner class of fishing vessel. Save for its length, it would have also met the requirements for the longliner class as defined by the Department of Fisheries. The only failing of the *Janet Louise* was that it did not meet the "political" requirements of the Federal Fishing Vessel Construction Assistance Program that were in place at the time of its launch; as a result, little was ever heard of Heisler's vessel.

During the course of my research, I compared the *Janet Louise* and the *David Pauline*, and the similarities between the two vessels became evident. At the time of its launch, the *David Pauline*, with the support of the Nova Scotia Department of Trade, Industry, and Commerce, was the first to meet all the requirements of the Federal Fishing Vessel Construction Assistance Program, some of which were not in force at the launch of the *Janet Louise*. There is some speculation among other builders that the design of the 1948 Youngs Island vessel served as the baseline for the Robar-built *David Pauline* launched in 1950. While there is considerable circumstantial evidence

In 1934, after the Dominion of Newfoundland gave up its claim to responsible government, six commissioners appointed by Great Britain governed Newfoundland. This body was known as the Commission of Government. In 1943, the commission began offering loans to fish companies willing to invest in fish plants and new vessels to expand and modernize the fishery. After Newfoundland joined Confederation in 1949, the federal government in co-operation with the new provincial Government of Newfoundland continued the practice along with the subsidy program that had been put in place for the Maritime fisher. At this time, Newfoundland did not have the know-how and the infrastructure to build fishing craft like those being constructed in Nova Scotia and therefore had to seek builders outside the province. Fisheries Products International would have been eligible for government-sponsored loans, but because the criteria for a federal subsidy at that time was for vessels with a minimum length of sixty feet, neither the FV *Zilch* nor the FV *Zoomer* would have qualified, despite their compliance with the federal standards for the longliner class.

to support this theory, there is nothing in the existing documentation that proves this to be so.

The FV *Janet Louise* was constructed for a well-known fisher, Captain Warren Levy, fishing out of Liverpool, NS. It is worthy to note that Captain Levy worked closely with Mr. Heisler in perfecting the design, and these discussions had a major impact on the construction of the subsequent vessels. Oddly enough, the remaining two vessels were not ordered for the Nova Scotia fishery but were part of a contract for three Cape Island–type fishing vessels placed by Fisheries Products International, St. John's, Newfoundland.

The *Janet Louise*, official number 179253, had a registered length of 50 feet, a breadth of 13.8 feet, and was powered by a General Motors 4071 Series diesel engine. What is of particular interest to this vessel is its design. First and foremost, from the half-model, it followed the basic design of the Cape Island snapper boat. Its profile showed the clear lines known to the Cape Island design, with the wheelhouse located forward and the accommodation space being

The FV *Janet Louise.* (FISHERIES COUNCIL OF CANADA)

forward of the wheelhouse with a trunk cabin to allow for better headroom. However, a study of the design of this vessel reveals some startling facts. Unlike the traditional snapper boat of similar size, the vessel was constructed with heavier and stronger materials. It had a separate shelter attached to and located aft of the wheelhouse. This shelter would come to be known as the "gurdy house" and was separated from the wheelhouse by a properly constructed bulkhead versus a single wall at deck level. The placement of this bulkhead and the gurdy house was one of the major design requirements in the Nova Scotia Department of Trade, Industry, and Commerce specification for the Cape Island–type longliner published in 1950, a year after the launch of the *Janet Louise.* What is even more startling is that the shelter/gurdy house was designed and constructed such that the port and starboard sides could be removed in fine weather. This feature, although rarely incorporated in longliners constructed after 1953, was again included in the provincial design specifications developed by Mr. Hines. Another feature was that the vessel was fully decked aft of the gurdy house and contained the required watertight bulkheads as specified by the Dominion Board of Steamship Inspection.

A comparison of the construction data of the *Janet Louise* and the *David Pauline* shows that with the exception of the *David Pauline* being five feet longer, all other parameters are identical.

The design and features incorporated in the FV *Janet Louise* were sufficient to have the Dominion Board of Steamship Inspection grant a certificate of seaworthiness as being part of the longliner class of fishing vessel. Given the vessel's length, at the time of its launch in early 1949 it would not have qualified for the Federal Fishing Vessel Construction Assistance Program; consequently, the detailed structural drawings that at this point were a part of the qualification criteria for the loan program would not have been required. However, there is little question that the provincial specification for the Cape Island–type longliner drafted by Mr. Hines and the provincial Department of Trade, Industry, and Commerce in late 1949 was surprisingly similar to the features that were incorporated into the *Janet Louise* that commenced construction in 1948 and was launched in early 1949.

The second vessel of interest was the FV *Zoomer*, official number 191226. It had a registered length 47 feet, a breadth of 13.6 feet, and a gross tonnage of 31.3. The construction details and features were very similar to those incorporated in the *Janet Louise*. The profile, wheelhouse, and living accommodations were similar, and like the *Janet Louise*, the vessel was decked aft, had a separate gurdy house aft of the wheelhouse, and was fitted with a mechanical gurdy and two watertight bulkheads, a design requirement that was stipulated by the Dominion Steamship Inspection Branch of the Canadian Department of Transport for wooden fishing vessels of this size of the longliner type. One bulkhead was placed just aft of the living quarters forward at the wheelhouse junction, and the second was placed between the gurdy house and the fish hold.

Newfoundland entered Confederation on March 31, 1949. The *Zoomer* was completed and launched in September 1949, and as a result, the Dominion Board of Steamship Inspection issued the certificate of seaworthiness. Given its tonnage, the vessel would have met the design and structural requirements for the longliner class that were in force at the time by the Canadian Department of Transport, and given its length it would have qualified for a federal subsidy had the 1951 amendment been in place. A review of the design details and final configuration showed that this vessel at time of launch would have also met the Nova Scotia construction standard for the Cape Island-type longliner that was published in 1950. The only requirement that is unclear in the registry documentation is the placement of watertight bulkheads, which is not recorded. The placement of the bulkheads was a critical structural requirement for a decked fishing vessel of this size and tonnage. The Dominion Board of Steamship Inspection would not have given the vessel a certificate of seaworthiness without them. Further research revealed that original registry documentation was completed by another party, separate from the Department of Transport. When Newfoundland joined Confederation, it took some time for the registry information to be transferred to the Canadian system, and during the transfer it was not uncommon for items in the original documentation to be omitted.

Unfortunately, in the 1947 to 1950 time frame, registry recording errors of this nature were commonplace in a number of classes of wooden fishing craft, due mainly to the requirement for additional data as a result of the implementation of more stringent design standards and reporting criteria. The installation of watertight bulkheads was unique to decked wooden fishing vessels over fifteen gross tons. It was also one of the critical requirements that separated the "government approved" longliners from the regular snapper boat that was defined by the federal government as being a "multipurpose" fishing vessel.

The third vessel of interest was the FV *Zilch*, official number 198238. The *Zilch* had a registered length of 48 feet, a breadth of 13.6 feet, and had a displacement of 21.17 tons.[60] Like the *Zoomer*, the *Zilch* was subject to the construction and safety standards laid down by the Department of Transport, Dominion Board of Steamship Inspection. A review of the registry information showed that this vessel was constructed with the required two watertight bulkheads, yet there is no record of the vessel being decked. There again appears to be a recording error, as the aft bulkhead would have only been installed if the vessel were decked aft. Another discrepancy appears to be in the registered tonnage. Given that both the *Zoomer* and the *Zilch* were almost identical in physical dimensions, the difference of ten gross tons seems questionable. With additional research and consultation with the Board of Steamship inspection, there was consensus that the individual that measured the *Zilch* used a different set of parameters for the calculation—a practice that was common at the time.

Despite these apparent discrepancies in the registry information, and although both the *Zoomer* and the *Zilch* were somewhat smaller in length than the standard longliner, like the *Janet Louise* both vessels met the same federal design criteria for the Cape Island–type longliner class as the *David Pauline* that was launched later, in 1950.

When the orders for the *Zoomer*, the *Zilch*, and a third vessel, the *Zanny*, were originally placed, Newfoundland had not yet joined Confederation. Therefore, the funding in the form of loans would have had to be sponsored by the Commission of Government for Newfoundland, and all approvals and requirements pertaining to the construction rested with that body. There is no evidence to suggest this changed after the colony joined Confederation in March 1949. Prior to Newfoundland's entry into Confederation, it would not have been able to take advantage of the Federal Fishing Vessel Construction Assistance Program, nor would it have been subject to the provisions of the Dominion Board of Steamship Inspection. At that time, therefore, there was no requirement to have approved design drawings forwarded to the Canadian government as was required by the 1947 amendment to the Federal Fishing Vessel Construction Assistance Program, nor did Newfoundland have the expertise to appraise the drawings. What the Newfoundland government and Fisheries Products International knew was that Mr. Heisler was a renowned builder with a reputation for quality vessel design and construction. Mr. Heisler's vessels were known for their strength and seakeeping abilities; therefore, the Government of Newfoundland did not the need Mr. Heisler to submit formal drawings. In addition, Mr. Heisler was well-known to the Nova Scotia Fishermen's Loan Board, and his designs were on file with their office. However, when Newfoundland became a Canadian province in March 1949, all three vessels were now to be registered in a "Canadian Port of Registry," and given their date of launch they would have been subject to Department of Transport construction and inspection standards. What is important here is they all met the Canadian federal standards.

A further review of Mr. Heisler's design showed that his vessels met the exact specifications and in some instances exceeded the standards that were specified by the Canadian Department of Transport, Steamship Inspection Branch and those that were developed and implemented by the Nova Scotia Department of Industry, Trade, and Commerce in 1950, one year *after* their launch and entry into the Newfoundland fishery, and that he did this one year *before* the launch of the *David Pauline*. There is, in fact, some question regarding the origin of the Nova Scotia Department of Trade, Industry, and Commerce design standard for the Cape Island–type longliner. The Heisler design and the features that were incorporated in his vessels match those included in the provincial specification, yet the design specification was published almost two years after the launch of the *Janet Louise*.

Between 1949 and the mid-1950s there were a number of fifty-five-foot wooden fishing vessels constructed in yards in Nova Scotia that looked identical or close to the *David Pauline*, and there has been an ongoing controversy amongst local fishers as to why these vessels did not meet the subsidy requirements and, in the context of the fishery, why they were not considered as being "government approved" longliners. While most if not all of these vessels did engage in longlining, there was a difference, and this difference was primarily due to the application of specific design and equipment standards to the construction of the vessel. To understand the difference, one needs a lesson in anatomy, which I am happy to provide.

CHAPTER 11

THE LONGLINER: A LESSON IN ANATOMY

*The more his equipment confines him, the
more will his catch depend on the inshore
runs of certain species, most of which
will be seasonal.*

– ROBERT MACGREGOR DAWSON, *REPORT OF THE ROYAL
COMMISSION ON PROVINCIAL DEVELOPMENT
AND REHABILITATION*, 1944

A DETAILED ANALYSIS AND CORRELATION OF DATA FROM THE
historical documentation held by federal and provincial agencies and
reputable private sources revealed that it was not until the construction of the FVs *Janet Louise* (1948) and *Zoomer* (1949) that the Cape
Island fishing boat design had evolved to the stage where the shore
fishery had a medium-sized vessel with the structural integrity for
fishing operations on the middle banks that at the same time was fully
compliant with the safety regulations for the longliner class of fishing
vessel as defined by the Dominion Board of Steamship Inspection.
It also revealed that despite their similar outward appearance, the
design and structure of the Cape Island–type, government-approved
longliner was very different from the normal forty-five to forty-eight-
foot Cape Island fishing boats that were engaged in longlining in the
near-to-shore fishery.

Between 1949 and the mid-1950s, there were a number of vessels
forty-five feet and above constructed in yards in Nova Scotia that

were based on the Cape Island model and outwardly looked similar in design to the *Janet Louise* and *David Pauline*. There was indeed a provincial construction standard called the General Specification Governing the Construction of Cape Island Boats that was in place before the provincial standard for the Cape Island–type longliner, and as the name would suggest, the specification was very broad in its scope. However, this document failed to address any specifics regarding how the vessel would be structured. From a provincial perspective, this specification was more applicable to the open Cape Island–type fishing craft that had a maximum length of forty-five feet and normally had a registered tonnage below fifteen gross tons. Also interesting is the discovery that in the late 1940s a large number of these vessels were registered as being below the fifteen-ton threshold when in fact their dimensions would indicate otherwise. The only explanation offered by present-day representatives from the Board of Steamship Inspection was that the number of changes taking place in the shipbuilding industry during this time, especially with respect to medium-sized fishing craft, meant that the calculations relative to the registry data were entirely dependent upon who measured the vessel and what data points were used. In the case of the Cape Island vessels, the drawings held by the provincial loan board appeared to have satisfied the federal authorities with respect to dimensions and calculated tonnage. There were seldom any questions raised by federal authorities. However, with the introduction of the Cape Island–type longliner and the Federal Fishing Vessel Construction Assistance Program's requirement to have all drawings prepared by a naval architect and approved by the Dominion Board of Steamship Inspection, there was suddenly more oversight and standardization of measurement methods, and adherence to the construction standards became more rigid and exacting.

Apparently, at the time of the promulgation of the provincial standard there was never any consideration given to the possibility that there would be any conflict with the federal standard. More importantly, it did not seem to be a consideration that the Cape Island–type snapper boat would evolve to become the Cape Island–type longliner. Did these smaller vessels engage in longlining? Absolutely. Some could be found fishing alongside the Cape Island–type longliner, but upon closer examination they lacked the strength, endurance, and hold capacity to effectively fish the middle banks to the same degree as could the government approved longliner. In fact, a good number pushed the design limits of the snapper boat to the point of jeopardizing safety, and sadly a number were lost. Although loans for construction of such vessels were granted by the Nova Scotia Fishermen's Loan Board during the entire life of the Cape Island–type longliner, the vessels that were constructed in accordance with the provincial specification at the time and failed to meet the criteria for the federal subsidy were classified as "multipurpose craft." It can be stated with considerable certainty that of the longliners constructed during the early '50s through to the late '60s, over 70 percent of the fishers that purchased Cape Island–type longliners had previously owned a Cape Island snapper boat in the range of forty-five to fifty feet in length. Of equal importance is that most of the snapper boats that were offered for sale and were replaced by the Cape Island–type longliner were less than ten years old. The answer was relatively straightforward to the question, why the change given the cost involved? Those fishers that purchased the Cape Island–type longliner made it very clear that they wanted a better vessel that could operate farther from shore and fish later into the fall and winter and do so with a greater degree of safety.

During my research, I investigated seven of the Cape Island snapper boats launched between 1950 and 1952 and classified at the time as being "multipurpose" to see if the reason they were rejected from the longliner class of fishing craft was politically based or whether there were solid structural/design reasons.[61] There is reason to believe that at least two of the vessels, the FVs *Swimm* and *Miss Margo II*, were rejected because the builder did not have formal federal government approval for their plans even though there is compelling evidence that the critical requirements of the structure were in line with the federal design specification. Much like the original design, the FV *Janet Louise*, they were technically compliant but failed the political requirements established by the federal Department of Fisheries for the subsidy that in fact defined the Cape Island–type longliner and therefore were not included in the list of government approved longliners. The remainder had design features that were not in keeping with the regulations applicable to the Cape Island–type, government approved longliners, and this was reflected in the overall cost of their construction.

Since 1949, there has been an ongoing controversy amongst fishers and others as to why the standard forty-five-foot vessels did not meet the subsidy requirements. A further controversy has existed as to why these vessels, which were engaged in longlining for ground-fish, were not considered by the governments of the time as being longliners—after all, that is what they were known as in their home ports.[62] The answer is both simple yet very complicated and defines the very body and soul of the wooden Cape Island–type longliner and separates it from other "Cape style" vessels.

For the past sixty years, there has been a misconception that the Cape Island–type longliner was "just another fishing vessel" and to some, just a larger version of the traditional Cape Island snapper boat. However, there was a difference—a difference that was directly related not only to the structural specifications that applied to a fishing craft of common dimensions, but also to a number of other specific factors that unless examined in detail will remain hidden when comparing both vessel types. The factors are as follows:

- the process that governed the loan and subsidy approval;
- the actual construction standards, process, and oversight;
- the introduction of newer propulsion technologies;
- the introduction of and increasing importance of hydraulics; and,
- the introduction of electronic technology in the form of voice radio communication, navigation aids, and fish finding equipment.

In comparing the Cape Island–type longliner to the conventional Cape Island fishing craft, especially those at or above forty-five feet in length, there was a significant difference in how the purchase and loan approvals were administered. In the case of the purchase of a newly constructed multipurpose vessel that was not receiving sub-sidy support, there was no clear definition as to what constituted a completed vessel. Even in the provincial specification for the Cape Island boat, reference is made to a number of options as to what the final configuration could be at time of delivery. Reference is also made to specific items such as the fitting of the wheelhouse and pro-pulsion system and provides some generic requirements depending on whether such systems were to be fitted by the builder or would be installed by the owner after delivery. That is to say, the vessel being constructed was not required to be delivered fully "ready for sea." Beginning in the mid-1940s it was not uncommon for a fisher to enter into a contract with a builder for the construction of a forty-five-foot Cape Island–type fishing craft with the builder supplying

A good graphic overview of the evolution of the Cape Island design from the small inshore craft to the larger snapper boat. The vessel on the far left appears to have a small dragger design that has been fitted with a gurdy house forward. The smaller vessels are inshore fishing craft that resemble the early Cape Island design. The vessel next to the wharf at the right appears to be a seiner. The white vessel in the middle is the FV *Oran*, which had an over-all length of forty-seven feet. Although this vessel was engaged in longlining and swordfishing and was known locally as a longliner, it was constructed to the provincial standard and was classified by the Dominion Board of Steamship Inspection as "multipurpose" and locally could be called a large "snapper boat." (NS ARCHIVES)

any variation from the bare bones completed hull up to and including a vessel ready to go to sea.[63] From the shore fisher's perspective, it was a matter of cost and what he was able to afford and/or finance. Another consideration was the type of groundfishing the fisher would be engaged in and during which seasons.

As a beginning step in the construction process, the fisher would approach a builder and obtain a price for the vessel he wished to have built. In some instances, the fisher would contact the Nova Scotia Fishermen's Loan Board and seek advice as to a suitable and approved builder for the type of vessel he was considering. In such cases the loan board would help the fisher find a suitable builder but would not act on behalf of the fisher during the initial negotiations. As part of the negotiations with the builder, the fisher could choose a number of options varying from the purchase of a completed hull up to and including a vessel that was in all respects ready for sea. Once the details of the construction and price were agreed upon by the fisher and the builder, the builder would provide a letter of commitment to the fisher stating that he would be willing to construct the vessel at the agreed upon price, outlining the level of work to be done and the configuration of the vessel upon delivery, and giving approximate dates for when the construction could be started and when the vessel would be delivered.

Armed with this information, the fisher would then make application to the Nova Scotia Fishermen's Loan Board for the required funding. Upon receipt of the application, the loan board would then contact the builder to confirm the details regarding price and design, and request that the builder provide a set of drawings lofted from the half-model for their review.[64] Once the application was approved, the loan board would prepare the required documentation in the form of a formal legal agreement between the loan board, the builder, and the fisher. The agreement outlined the responsibilities of each party to the contract, acknowledgement of the fisher's deposit, schedule of payment and particulars, and an attached schedule that detailed the work and configuration of the completed vessel.

The process applicable to the purchase of the Cape Island–type longliner basically followed the same path up to the point of loan approval except that during the initial loan granting process, the fisher would have to make application for financial assistance from the Federal Fishing Vessel Construction Assistance Program. As a result, there were some major differences in the content and format of the formal loan agreement.[65] First and foremost, the clauses relative to the approval of drawings and specifications were amended to reflect that the approval authority for the drawings was no longer the Nova Scotia Fishermen's Loan Board but the federal Department of Transport and in particular the Dominion Board of Steamship Inspection. Another point to note is that unlike the original provincial loan application, the one for the "government approved" longliners had an additional section that made direct reference to order-in-council P.C. 2490—the Federal Fishing Vessel Construction Assistance Program. This section detailed the specific requirements and responsibilities of the fisher with regard to the granting of the loan and subsidy.

In addition, a legal schedule attached to the agreement explicitly outlined that the vessel had to be delivered to the loan board and be "in all respects ready for sea" with exception of bedding, messing, and fishing gear. It further detailed all the fittings relative to the structure and power plant, instrumentation, rigging, safety equipment (including one dory), firefighting apparatus, and electronic equipment if so fitted.[66] Every requirement had to satisfy the Board of Steamship Inspection—a direct reference to the federal requirement for the wooden fishing vessels of this class—and the provincial design specification for the Cape Island–type longliner. Prior to being granted a certificate of seaworthiness, the vessel would have to complete a rigid set of compliance/builder's trials. Certificates would be granted for a set period of time, after which the vessel would be docked and subject to a structural and machinery inspection prior to the certificate being renewed. When all the requirements and regulations were satisfied, the loan board would disburse the final funds to the builder, and the vessel would be released to the ownership of the purchaser. It is worthy to note that the loan agreement also outlined the builder's responsibility and established a fixed delivery date. In many instances the builder treated the construction of each hull as an individual project and established timelines for the completion of various activities. A copy of such an agreement, with personal information removed, is included herein as Appendix B.

From a pure political and regulatory perspective, the line that divides the Cape Island snapper boat and the longliner can be found in reviewing the construction standards and requirements embedded in the federal regulations implemented in the 1945 time frame and the 1947 amendment to the Federal Fishing Vessel Construction Assistance Program. The standards for the "longliner class of fishing vessel" enforced at the time by the federal Departments of Transport and Fisheries demanded that each builder wishing to engage in the construction of the longliner had to fully satisfy all their structural and design criteria from the submission of formal plans to the completion of government supervised builders' trials. There was also a need for the provincial loan board to have a detailed specification that would not only define the Cape Island–type longliner as a separate class of fishing vessel, but also would integrate the structural and design standards of the Dominion Board of Steamship Inspection. The bottom line of all this was that if the vessel qualified for assistance under the Federal Fishing Vessel Construction Assistance Program, then it was deemed to have satisfied all the necessary federal and provincial regulatory requirements and was classified as a longliner.

Another more practical reason for this delineation was that just prior to 1949 and onwards, there was a large number of the forty-five-foot and larger Cape Island–type snapper boats constructed in Nova Scotia yards that were delivered and remained open construction. The majority, depending on the geographic area, engaged in the lobster fishery and had hatch-type deck coverings fitted aft of the wheelhouse for groundfishing, and as such they were considered by the Dominion Board of Steamship Inspection as being open "multipurpose" fishing craft. Thus, they were not subject to the same regulations as those applied to the longliners seeking subsidy support. A quick review of the cost of these vessels compared to the longliner shows the longliner to be much more expensive. This again was due to the type and amount of material and equipment used/fitted in the longliner compared to the conventional snapper boat. As an example, in 1951 the average cost of a snapper boat was in the order of $5,000 compared to $17,000 for vessels like the *David Pauline.* The province had the authority to grant loans for such conventional snapper boat–type vessels, and construction of such craft continued well into the 1980s. Although some of these vessels fished the same grounds as the longliner, the majority remained engaged in the near-shore fishery and fished in the spring, summer, and early fall. Their annual days at sea were far fewer than the Cape Island–type longliner. The two classes of fishing craft complemented each other in the shore fishery and proved to be a very viable and effective economic "team," especially for the smaller fishing communities that had a small to medium-sized fish plant.

A review of the design requirements for the Cape Island–style longliner shows some very subtle and unique construction requirements when compared to those applied to vessels that were not subject to

A Cape Island–type longliner under construction at the Harley Cox and Sons yard in Shelburne, NS. (NS ARCHIVES)

subsidy approval. The average length for the first Cape Island–type longliners was around 57 feet, and the beam varied between 14.5 feet and 17 feet depending upon the builder and the model being used. In the mid-1950s and continuing into the 1980s, a forty-eight-foot version of the vessel gained popularity. However, when the forty-eight-foot version started to appear, the building standards and structural requirements did not change, and the same material specifications remained in force for these smaller versions as was for the larger vessels.

Highlighting some of the regulations and requirements that were applicable specifically to the longliner demonstrates some of the more subtle differences between the longliner and other fishing craft. In keeping with the requirements of the provincial construction standards, the vessels had to be constructed in a covered facility that was weathertight.[67] Native wood was used throughout the construction process.[68]

The keel and stem were normally eight-by-nine-inch birch with a four-by-one-inch yellow birch hog[69] completing this principal member. In most cases these structural members were larger and heavier than those found in the conventional forty-five-foot snapper boat. Frames were normally two-and-a-quarter-by-two-inch steamed grey oak or juniper, depending upon the builder, and bent on eight-inch centres. In some instances, again depending upon the builder, these frames measured three-by-one inches and were double-bent eight-inch centres. The vessels of this size were planked with spruce that had a minimum thickness of one and a half inches and a maximum width of eight inches.[70] An additional sheathing of a half-inch-thick oak or suitable hardwood was fixed to the main plank commencing at the stem post forward at the waterline and extending from the joint of the rabbet at the keel forward at the stem post and continuing aft along the counter to the stern.

This is in stark contrast to the dimensions of the materials that were used in the construction of a forty- to forty-five-foot Cape Island snapper boat. On this type of fishing craft, the thickness of the plank had a maximum of one inch with the timbers/frames being one-by-two inches bent on eight-inch centres. In the Cape Island longliner, the bilge stringers—one of the main longitudinal structural supports for the vessel, which extended from the waterline forward to the stern—were normally two-by-six-inch spruce. In some vessels it was not uncommon to install two such stringers for added strength. These were bolted through the frames and plank at predetermined intervals. Spruce was also employed for all structural members such as internal clamps and shelves. Deck beams were made of four-by-seven-inch birch with three-by-seven-inch auxiliary beams of oak or other suitable hardwood covered with one-and-a-half-by-two-inch spruce decking.[71] Each of the beams was reinforced with knees that followed the natural curvature of the tree root, and these were bolted through the structural members.

One of the more unique design requirements of the Cape Island–style longliner that has gone virtually unnoticed during its life is the fact that the whole vessel was divided into a minimum of three watertight compartments: the forecastle, the engine room, and the fish hold. Two watertight bulkheads that extended from the main deck to the top of the keel proper separated the three compartments. The federal construction standards of the time provided considerable detail as to how these bulkheads were to be constructed, installed,

Inside the *Oran II*, one of the original Cape Island–type long-liners launched from the John McLean and Sons yard in 1957. This shows the placement of the double-bent frames, the spacing, and the elements of the adjoining structure. (AUTHOR'S COLLECTION)

and sealed. In the case of the bulkhead aft of the engine room, the regulations specified that the propeller shaft(s) was to pass through the bulkhead via a watertight gland. The installation of this bulkhead and gland by regulation was to be inspected "as soon as practical" after installation. In addition, the integrity of the bulkheads and throughway for the shaft(s) was to be subject to examination by federal inspectors as part of any future inspection conducted by the Dominion Board of Steamship Inspection. Each of the compartments formed by the watertight bulkheads were to have a capability for evacuating water from the space. Therefore, suction sumps and appropriate plumbing had to be installed in each compartment and connected to the main bilge pump.

In the engine room looking at the forward watertight bulkhead on the Cape Island–type longliner *Oran II*. (AUTHOR'S COLLECTION)

The forecastle was arranged for a crew of five. Entrance to the accommodation space was gained from the wheelhouse located in the deckhouse directly aft of the forecastle. The crew's quarters had to be equipped with suitable bunks, storage lockers for personal gear, a stove, dining table, cupboard, and lockers for the storage of provisions. The crew area also had to have a sink and running potable water. All these facilities were subject to government inspection.

The depth of the Cape Island–type longliner allowed for a small engine room. The engine room was located directly below the deckhouse and was subdivided into an engine space and a fuel tank compartment with the engine being in the forward end of the space. A non-watertight bulkhead separated the fuel tank compartment and engine room. The engine room contained the engine-driven pumps used for de-watering the bilge and the watertight compartments, firefighting, and washing down the work area. This space also housed the main battery bank and the auxiliary equipment such as emergency generators and power supplies to support specific navigational aids.[72] The fish hold was directly abaft of the fuel tank compartment and was accessible to the main deck by a raised hatch that could be made secure and watertight while at sea.

The deckhouse, as was common with the Cape Island–style design, was situated approximately amidships. It consisted of the wheelhouse forward and a closed-in space directly aft of and separated from the wheelhouse by a bulkhead and door to prevent water and fish offall from entering the wheelhouse. On the official drawings and in government specifications, this space was commonly called the gurdy house but was known by most fishers as the "slaughterhouse." This space was designed for the protection of the crew in inclement weather, and it was in this working space that the trawl hauling gear or gurdy was located along with a number of built-in fish pens that allowed for the dressing of fish before being placed in the fish hold proper. Accommodation was made in this space to allow for the operation of engine controls and steering. A small sliding door afforded access to the fish hold below the after deck. All decks and floors were to be made watertight and scuppered.

The federal documentation relative to the minimum design requirements for wooden fishing craft also specified the minimum power required for a given class of vessel of a specified length and tonnage. These requirements are further divided into specific power levels for specific reduction gearbox requirements. For example, a vessel that is operating on a 2:1 reduction of engine RPM to shaft RPM would require more horsepower than one operating with a 3:1 reduction ratio. The provincial design specification approved by the Department of Trade, Industry, and Commerce in 1951 takes this one step further. In this specification, the province stipulates that only marine engines approved by their marine branch could be used and even goes so far as to name engine types and manufacturers that would meet their approval. This practice would be considered very inappropriate in today's marketplace. Regardless of the optics of such a statement, such actions were deemed acceptable at the time, and there is no record of any one manufacturer questioning the government's action. The only federal provisions regarding propulsion requirements were the required horsepower as per the federal design standard for a specific class of vessel and the reduction gearbox installed; no reference is made to specific engine manufacturers.

This raises a very interesting point and further separates the Cape Island–type longliner from the snapper boat. In the case of the standard Cape Island–type fishing vessel, the provincial guidelines for construction do not specify any particular propulsion system. Although a good number of these vessels were powered by either gasoline or diesel marine engines, there were a number that used automobile engines fitted with a marine clutch and transmission. In the 1950s and even into the early 1960s, a large number of Nova Scotia fishing vessels used in the lobster and coastal groundfishery were still using automotive gasoline engines. By 1952, longliners were coming off the ways with diesel engines manufactured by Leyland Acadia, General Motors, Rolls Royce, Caterpillar, and Cummins to name a few, with no question being raised by the provincial loan board. With the passage of time, smaller versions of these engines were finding their way into the snapper boat fleet.

The technological advancements that came out of the Second World War relative to target detection, range finding, and passive electronic navigation soon found application in the commercial fishing industry. These systems, as rudimentary as they may seem when compared to today's equipment and systems integration, added a new dimension to the safety and capability of the medium- to large-sized fishing craft. Wireless radio systems that had their beginnings in merchant vessels and warships in the early twentieth century evolved from Morse code transmission into long-, medium-, and short-range ship-to-ship and ship-to-shore voice communication. With the major advances made in the field of powered flight and aviation during the Second World War, similar technology was beginning to be applied to aircraft. However, before this could be made possible, science had to devise systems that were smaller, lighter, and could accommodate more efficient power sources that would allow the system to be installed in small and medium-sized aircraft. The imperatives of the war effort forced these technical issues to be resolved quickly, and by 1949 hybrids of the airborne systems were being adapted and installed in smaller vessels such as trawlers, draggers, and fishing craft such as the Cape Island–type longliner. By 1953, all longliners being built in Nova Scotia were equipped with some form of ship-to-shore voice communication. As radio technology advanced from vacuum tubes to solid-state and integrated circuitry, radios became smaller and more efficient and by the mid-1980s could be found installed in fishing vessels of all types and sizes. However, in the early days of the longliner fishery, the capability to have a radio on board was considered a major technological advancement.

In the realm of passive navigation, systems such as Loran-A were developed that allowed vessels with a specific receiver installed to intercept signals from base and satellite stations ashore and in doing so provide the captain/navigator an effective means to correlate the ship's position. The system, whose name derives from "long range navigation," was designed primarily for open ocean navigation and decreased the reliance on celestial navigation, especially during periods of poor celestial visibility, and was rapidly brought into service during the Second World War. Again, the systems found their way into aircraft, and it is not surprising that after the war, surplus airborne systems such as the APN 4 and APN 9 (the initialism APN stands for "aircraft position navigation") were installed as a primary navigation aid in trawlers, draggers, and upon their introduction, the longliners. Depending upon the atmospheric conditions and the experience of the operator, the early loran systems provided a reasonable degree of accuracy for open ocean navigation. As the vessel came closer to shore, the system proved to have less accuracy, mainly due to the interference of landmasses with the propagation of the various chain transmissions. In the case of the Cape Island–type longliner, loran was seen as a major asset, as the ship's master had a more effective method of not only plotting the vessel's position, but in some cases was able to plot the location of fishing gear. Although the system was less accurate the closer one came to the shoreline, it was a major improvement over using only the magnetic compass and dead reckoning during periods when celestial navigation was not practical. In the case of the Cape Island–type longliners, most of the fishers who owned and operated these vessels had very little open ocean navigational skills save for using the magnetic compass. Because they were primarily involved in the near-to-shore or coastal fishery, there was no real requirement for celestial navigation. The introduction of Loran-A filled this gap in navigation capability and proved to be a very useful tool. The Loran-A system remained in service worldwide and on all types of vessels well into the 1970s.[73]

In the late stages of the First World War, British and Canadian scientists developed a sound-based underwater detection system known as ASDIC. This technology was developed out of the need for a method to detect enemy U-boats. It was discovered that by directing a sound signal into the water and having its echo bounce back to the vessel and correlating the time and distance of the signal electronically, the object could not only be detected, but its range/depth could be determined. This detection system became the forerunner of the depth sounder and fish finder that were to play a prominent role in the trawlers, draggers, and longliners. From this rudimentary beginning, this system, like others, underwent a number of major advancements in a very short period of time. By the end of 1945, this detection system now known as sonar ("sound navigation and ranging") had evolved into echo sounders that could determine the depth of water beneath the vessel's hull and, with the addition of a chart-type recorder, could provide a graphic display of the ocean's bottom, its depth, as well as any object including schools of fish between the bottom of the keel and the ocean floor. By 1949 sounder systems were being installed in the larger fishing vessels, and by 1952 they were finding their way into the wheelhouse of the longliner. It did not take fishers long to realize that with practice and a good understanding of the capabilities of their system, the sounder itself was a very viable and effective navigational aid when operating in coastal waters.

Another system that evolved from the Second World War was radio detection. This system operated on the principle of receiving radio signals from a known source or frequency bandwidth and determining the direction of the signal relative to the position of the ship and/or aircraft receiving the signal. Originally this technology was used to intercept submarine and ship-generated radio traffic in an effort to determine the location of enemy units. It was used both at sea and by patrol aircraft and proved quite effective. As the technology improved, the system was able to detect and locate shore-based voice and carrier-wave radio traffic.

It's not surprising that this technology found its way into merchant vessels, warships, and aircraft as a navigational aid. By the end of the Second World War, navigational charts, especially those in North America, contained prominent features such as radio transmitter towers and radio beacons. Supplementary publications such as navigational pilots started to include the radio frequencies of prominent beacons and towers, as well as a location, and frequencies of local commercial radio stations. A vessel at sea could, depending upon the frequency and the range assigned to the tower, locate a specific beacon and correlate the ship's position relative to the transmission site. The system worked very well for most applications, but it was restricted to the range and frequency of the tower being queried and the availability of similar towers/beacons. The system, although valuable, was further restricted when used close to shore, as depending upon the topography of the land, the land mass itself could interfere with beacons and towers. In the case of Nova Scotia, the number of radio towers close to the smaller coastal fishing ports was very limited, but the system proved most useful nonetheless. Those that used this system in the Cape Island longliners affirmed that it did provide directional information in periods of poor visibility and was fairly reliable in assisting the captain plot the vessel's position and mean line of approach to a specific port or harbour when used in conjunction with other navigational aids. It proved more valuable in determining an approach to major centres such as Halifax, Sydney, or Digby where a host of radio towers were located.

Perhaps one of the most useful technologies to come out of the war was the invention of radar ("radio detection and ranging"). The basic principle of radar was the transmission of a pulsed radio signal in a specific direction and the detection of the returning signal from the radio wave contacting an object. By the use of electronic measurement, the operator can determine the location of the object, its range, and whether the object is stationary or moving. From its rudimentary beginnings in 1939 as a ground-based detection system, radar soon found application at sea and in the air as a target detection system that expanded and evolved into a very effective navigational aid and collision avoidance system. In the beginning, the primary use for the system was target detection, with the signal being provided by fixed or hand-operated antennae. Any returned signal was displayed on a calibrated yet very basic cathode ray tube graphic display. A major breakthrough came about with the introduction of an automated rotating antenna and a receiver fitted with a more advanced cathode ray tube[74] that provided 360-degree visual coverage known as a "planned position indicator" or "scope." The radar system was now capable of not just detecting objects that entered its effective range but could display landforms and other targets at night and in dense weather formations with a complete 360-degree radius of coverage from the transmitting facility.

One of the major drawbacks in the rapid advance of radar was the size and weight of the power system required to run it. By 1942, this problem was coming to some resolution when the Royal Air Force developed a radar system that could be fitted to medium and large aircraft. These systems proved very effective and were subject to considerable technological advancement by war's end. By 1944, the advancements made by the RAF with airborne radar were being exploited by the Allied navies and the merchant marine. There was a requirement for a smaller and more compact system that could be fitted to smaller trawler-size[75] vessels and patrol craft. The result of their efforts was the development of a radar system that could use a 32-volt power system that was compact enough to be fitted to medium-sized ships and coastal vessels. The only major drawback was that the reduced power requirements also reduced the effective range to approximately twenty nautical miles. This range reduction was of no great consequence for commercial marine operations. It wasn't long before the major manufacturers such as Marconi, Decca, and RCA began a major push to have their systems installed on fishing craft.

Radar systems were a new technology to the fishery and were relatively expensive. By the beginning of the 1950s, radar systems started to appear on most large trawlers and some of the smaller draggers. Radar did not start to appear in the longliner fleet until 1957.[76] By the end of 1958, most longliners built previous to this date were retrofitted with a radar system, and all built after that date were radar equipped. In its day, radar became one of the most useful pieces of electronic equipment to be fitted to a fishing vessel.[77] Radar allowed the vessel to have a "set of eyes" in conditions of fog or natural darkness. With experience, longliner captains were able to mark and locate their trawl lines more effectively than by visual means, were made more aware of the sea space around them with respect to other marine traffic and collision avoidance, and could more safely navigate in and out of harbours in fog or darkness. It wasn't until the mid-1970s and the introduction of solid-state circuitry that radar receivers had advanced to the stage where they could be fitted to almost any size of fishing vessel. Until the introduction of global positioning systems (GPS) in the 1990s, radar remained one of the prime electronic navigation aids of the fishing industry.

In the early 1960s another navigational aid, the Decca Navigator, started to appear on the larger fishing vessels. Originally developed for the Royal Navy when there was a need for a system that could be used for accurate landings on enemy coastlines, the Decca Navigator was further developed after the war for the merchant service in and around the United Kingdom and European waters. By the mid-1950s fishing vessels were becoming the primary users of the system; although it did not have the open ocean range of loran, Decca was offering much better accuracy in the 60 to 120 nautical mile range from land and in coastal waters.[78] By the mid-1950s and onward, Decca Navigator systems were being installed in most large and medium-sized trawlers and draggers. In most of these vessels, loran was still part of the navigational suite. The operating principle of the Decca system was similar to that of loran. The system consisted of a number of land-based radio beacons organized into chains. Each chain consisted of a master and three "satellites," coded as red, green, and purple. Each station transmitted a continuous signal. By comparing the phase difference of signals from the master and one of the satellites, a set of hyperbolic lines of position would be produced. The lines of position were marked in the appropriate colour on the chart. The on-board receiver identified which lines the vessel was located on or near, and by plotting the intersection of the lines of position, the ship's position could be plotted. The method used by the Decca system was considered more "user-friendly" than loran. This feature of the Decca Navigator made it quite popular with fishers, and by the early 1960s this navigation aid was starting to appear in a number of Cape Island–type longliners. In the mid-1970s both loran and the Decca Navigator were rapidly being replaced by a more accurate system known as Loran-C. The vacuum tube technology of the original loran systems was becoming outdated and expensive to support not only for owners of receivers, but also for the actual stations that transmitted the signals.

Advancements in solid-state electronic systems, component miniaturization, and system automation were rapidly replacing the tube technology systems inherent in the APN 9, loran, and early radar system. Loran-C was very cost effective to establish and operate, was more accurate and user-friendly than the earlier loran and Decca systems, and the receivers were very affordable to the average fisher. It is not surprising that by the early 1980s Loran-C became the most widely used navigational system in the world and was rapidly being installed in vessels of all types, including the Cape Island–type longliner.[79]

It is clear that the Cape Island–type longliner was not just "another fishing vessel" or a "super" snapper boat. Its design, construction, and imbedded technology gave the longliner its "heart and soul." By today's standards, such a fishing platform would be considered very crude, but to the fishers of the 1950s and early '60s it was a significant advancement. The specifics of the design and the oversight of construction standards coupled with improved navigation, communication, and fish-finding aids were a source of pride and accomplishment for the builders and those that supported the introduction of the Cape Island–type longliner to the fishing fleet. These proud Nova Scotians had accomplished a remarkable goal that a few years before seemed a faraway dream. By the mid-1950s the Cape Island–type longliner was not just a fishing vessel, it was in essence a fishing system that embodied all the requirements that were lacking by any previous fishing platform in the shore fishery. It became a recognizable icon for the shore fishery in much the same way as the large steel trawler did for the banks fishery and the mid-size dragger for the scallop industry. Yes, it had a heart and yes, it had a soul—a soul that remains in the memory of those who built and sailed this fine Nova Scotia craft.

CHAPTER 12

THE OTHER LONGLINERS

The fisheries present another magnificent opportunity for the Dominion government to revive the economy of the Maritime Provinces and especially Nova Scotia.

- THE HONOURABLE C. D. HOWE, QUOTED IN THE *GLOBE AND MAIL*, OCTOBER 13, 1944

THE PRIMARY FOCUS OF THIS BOOK IS THE SHORE FISHERY AND THE role Cape Island–type longliners played in its recovery and expansion. However, one would be remiss if recognition was not given to the fact that there were other hull designs that met the same federal technical standards required to obtain subsidy support under the Federal Fishing Vessel Construction Assistance Program and were classified as a longliner. Unfortunately, despite their ability to actively participate in the ground and swordfishery as longliners, by the late 1960s into the mid-1970s, the majority had been modified to operate as a trawler/dragger.

As early as 1942, Nova Scotia, in consort with the federal government and with its financial backing, had started to examine and fund the construction of medium-sized wooden trawlers. Although restrictions on trawlers had been set in place ten years earlier, the two levels of government saw the need to increase fish production to support the war effort and proceeded without any formal government repeal of the restrictions. What is ironic is that the Nova Scotia government had been fighting since 1936 for funding for the shore fishers to have better fishing craft but by 1942 had accepted without question or concern for the shore fishers the federal government's proposal to subsidize the construction of draggers. What is also interesting is that there was no mention of any support for the construction of longliners, small or medium. From a shore fishery perspective, the biggest flaw was that the minimum size allowed for granting of the subsidy was seventy-two feet, too large and costly for the average shore fisher. This did not change until 1950.

In December 1943, the wooden dragger *Sea Nymph* was launched at the LeBlanc Shipyard, Weymouth, NS, while two others were under construction at the Meteghan Shipbuilding Company, Meteghan, NS. When the trawler restrictions were officially lifted in late 1944, there was an initial push starting in New Brunswick to build small and medium-sized trawlers that would meet the subsidy requirements in place at the time. The same dragger hull design was found to be quite suitable for the herring fishery; this coupled with the success being enjoyed by similar craft from ports along the coast of Maine and Massachusetts raised the interest of fishers and fish processing plants in New Brunswick. The hulls were based on three basic designs: the Gloucester, Chaleur, and Charlotte-type trawlers/draggers that were initially designed by well-known naval architect firms such as Eldridge-McInnis Inc. (Hingham, MA) and Potter, MacArthur, & Gilbert (Boston, MA), to name a few. Beginning in 1943, these hull designs met the building and structural requirements laid down by the American Bureau of Shipping and had drawings that were approved by recognized naval architects. Therefore, from a Canadian government perspective, such designs would easily qualify for the financial support of the Federal Fishing Vessel Construction Assistance Program, meet the formal approval of the Dominion Board of Steamship Inspection, and allow the individual provincial fishers' loan boards to have easy access to approved drawings that would satisfy the federal Department of Fisheries.

Although the formal legislation to lift the Canadian ban on trawlers was passed into law in 1944, the amendments to the subsidy legislation passed by Parliament that same year would not allow an individual fisher to be granted the subsidy and own such a fishing craft. It is not surprising, therefore, that the majority of the vessels constructed were for medium to large-sized fish dealers and investors. Beginning in the 1946–47 time frame, the fishers in the North Shore area of New Brunswick who were more disposed to groundfishing in the Gulf of St. Lawrence found that with minor modifications such as the addition of a gurdy house amidships and the installation of a mechanical trawl hauler, the hull designs could meet the subsidy criteria for being called a longliner.

In Nova Scotia, the inshore fishers who already believed that the larger fish dealers were squeezing them out were very critical of the trawlers that they had dubbed "sea scavengers." The limitations placed on the subsidy by the 1944 amendment further increased their militancy, and as a result they paid little attention to the modifications that were occurring to the trawler design. The construction of the FV *Sea Nymph* in 1943 and the additional two hulls constructed at Meteghan caused a migration of sorts of the three hull designs to the Nova Scotia side of the Bay of Fundy. The Nova Scotia yards with the capability to construct this hull form were receiving orders mostly for the New Brunswick and Prince Edward Island markets. It was not long before other Nova Scotia yards started to develop their own hull designs using the original Maine/New Brunswick hulls as the baseline. During the same time frame, some of the New Brunswick-designed medium-sized trawlers that were built as longliners started to appear in Nova Scotia ports, and the fishers themselves began to see the advantages that this design held.

The reasons for fishers initially choosing such designs over the Cape Island–type vessel are varied. First and foremost, in 1945 through until 1950, the Cape Island design for the longliner had yet to be formally approved. Notwithstanding the implementation of regulated building standards, the majority of the Cape Island–type fishing vessels of the time were too small to qualify for the federal subsidy. A further impediment to adapting the Cape Island design was again the 1944 subsidy regulations that would not allow the granting of the construction subsidy to singular ownership. Therefore, a good number of Nova Scotia fishers continued to engage in longlining with the forty- to forty-five-foot, standard Cape Island–type snapper boat, which although an excellent craft, did not have the endurance and capability to fish the middle banks that later became known by some as the "near-shore fishery." Along the Eastern Seaboard of Nova Scotia, the major fish dealers that had been involved in the Grand Banks fishery of the 1920s through to the outbreak of the Second World War were more interested in the large steam trawlers than the small wooden trawlers, and very little attention was given to the use of this hull design as a longlining platform.

Another factor was the ability of the Nova Scotia yards to construct this type of vessel. Beginning in 1945, the shipyards that had the capacity to build forty-five-to-sixty-five-foot fishing vessels had varying capabilities. Some yards, predominately on the east coast of Nova Scotia, had experience in building the Cape Island–type vessels up to fifty feet and, of course, had considerable capability in building schooners and other sailing craft. However, from 1947 through to the early 1950s, the actual experience in building the Cape Island–type longliner to the standards required was initially limited to a few yards. In southwestern Nova Scotia and in the Bay of Fundy region, the close proximity of the shipyards in New Brunswick and northern Maine had allowed the medium to large builders to enter the market being created by the demand for the medium and small trawlers and as a result were gaining valuable experience building small dragger-type hulls. Foremost among these Nova Scotia builders were A. F. Thériault and Sons Ltd., Meteghan River, NS, and the Wagstaff and Hatfield Yard in Port Greville, NS, both located on the Bay of Fundy. It is not surprising that, in 1950, when the subsidy legislation was amended to allow singular ownership, a small number of fishers eager to take advantage of the financial support that was available to obtain larger vessels elected to purchase a longliner with no real preference as to the hull type. Some fishers chose alternate designs based on personal preference and a belief that the schooner and/or dragger-type hull with the wheelhouse aft had better seakeeping abilities than the Cape Island type. Others looked at the designs as having the potential to be reconfigured to other fisheries, such as dragging, should the need arise. This was particularly applicable in the early stages of the subsidy program when the majority of construction was for the medium-sized fish dealers and groups of fishers that met the purchase requirements for the granting of the subsidy.

Commencing in 1949 and continuing into the 1960s, the Nova Scotia shipyards both on the Atlantic side of Nova Scotia and in the Bay of Fundy built a large number of dragger/trawler-type vessels. One must remember that the subsidy was open to draggers and seiners and, given that the herring and scallop fishery was very important to fishers on the Bay of Fundy, it is not surprising that the demand for these vessels was continually on the rise. Over this period each builder began to develop their own unique hull forms, and the individual vessel design, like the Cape Island–type longliner, became quite identifiable. The size of the vessels varied to match the length requirements of the subsidy in force at the time. There is no question that the number of small and medium-sized trawlers/draggers built between 1949 and 1985, the period of this study, exceeded the number of Cape Island longliners being constructed; however, the number of trawler/draggers that were built for and actively involved in longlining was significantly less than the number of Cape Island–type longliners engaged in that activity.

One of the first of these alternately designed vessels was the FV *Herbert R. Swim*, launched in 1951 from the W. C. MacKay and Sons yard in Shelburne, NS. It was built for the fishing firm of Swim Brothers Ltd., Yarmouth, NS, and as it was to be owned by a company, qualified for the Federal Fishing Vessel Construction Assistance Program.

Larger than the Cape Island–type longliner, this vessel was somewhat unique. Its hull followed the basic schooner design but had a wider beam than the conventional schooner hull and had a rounded stern. Unlike other vessels of this design, this vessel had its wheelhouse built amidships and had workspace both forward and aft of the wheelhouse. It is unclear what classification it received to satisfy the subsidy requirements, whether dragger or longliner, but it was built with the capacity and capability to be converted from a dragger to a longliner or vice versa. What is known is that it began its life on the fishing banks as a longliner primarily engaged in the halibut fishery, converting to swordfishing during the summer season.

In 1952 another longliner that again had its original design based on the Maine/New Brunswick vessels started to appear in Nova Scotia ports. At first glance its hull form was identical to the sixty-foot draggers/side trawlers that were being constructed in yards along the South Shore and Bay of Fundy area of Nova Scotia. Like the "double end" design, these vessels were modified to meet the criteria for longlining by the addition of a gurdy house just aft of the forecastle and the installation of a mechanical trawl hauler. The design had a wider beam, and over time each builder changed the design to satisfy the needs of the customer. These modifications did not compromise the standards set by the Dominion Board of Steamship Inspection and as such had no difficulty in qualifying for provincial loan funding and the federal subsidy. As with the other similarly designed vessels, they proved to be a successful longlining platform; however, like

The FV *Herbert R. Swim.* (FISHERIES COUNCIL OF CANADA)

their fellow vessels, the majority were converted to trawling/dragging operations primarily being employed in the growing scallop fishery.

In 1953, yet another longliner design was introduced in the Nova Scotia shore fishery, this time based on the Maine/New Brunswick "peapod" or what was known in Nova Scotia as the double ender–type herring seiner/trawler. The vessel had a principal dimension of 60 feet, a breadth of 16.5 feet, and mean depth of 6.6 feet. As the basic design was already known and approved, the structure of the vessel met the requirements and specifications of the Dominion Board of Steamship Inspection and as such qualified for the Federal Fishing Vessel Construction Assistance Program. From a provincial perspective, this type of hull design, like others of similar design, was well known, and a standard specification document was already in existence and satisfactory to the federal Department of Fisheries. Therefore, there was no difficulty for the Nova Scotia Fishermen's Loan Board to grant funding for such a vessel while at the same time satisfying the federal subsidy requirements. In keeping with the original design, the wheelhouse was located aft, and below the wheelhouse was the engine room. The engine room was separated from the forward section of the vessel by a watertight bulkhead that extended from the keel to the main deck. The crew's quarters were located in the forward part of the vessel with the fish hold again separated from the aft part of the vessel by another watertight bulkhead. These bulkheads were a critical requirement of the Dominion Board of Steamship Inspection for a vessel of this length and tonnage. The work area was located on the main deck between the fo'c'sle and after wheelhouse. The original design was modified by Mr. MacKay to accommodate a wider beam, but it is unclear if there were any other major modifications made. To satisfy the requirement for a separate gurdy house, a structure was placed just aft of the forecastle area and forward of the main fish hold.

The first vessel of this design to classify as a longliner was the FV *C Ola J*, built for Capt. Cecil Payzant of Port Mouton, NS. Following its launch, a number of others identical in design were completed. From all accounts, the vessels performed quite well and participated in the shore fishery for a number of years. In the 1960s, a number of these small vessels were converted to scallop draggers where they remained until their removal from service. A number of those reconfigured to dragging operations were replaced by the Cape Island–type longliner.

Another design that cannot be forgotten was the schooner. By the beginning of the Second World War and with the introduction of the internal combustion engine, the days of the Grand Banks schooner and the dory fishery were slowly fading into history. By 1942, very few schooners remained under sail. The large schooner fleets that had sailed from ports such as Lunenburg had steadily declined in number and continued to do so. Some, like the famous *Bluenose*, had been sold to offshore interests for use as cargo vessels, while others had been sold to Newfoundland interests. Those that remained had the topmast removed, the sails put away, and had engines installed. When the subsidy program became available to the fishers of Nova Scotia, one of the provisions allowed for a subsidy to be granted for the conversion of schooners to powered longliners. Although the average age of the remaining schooners was approaching twenty-five years, a small number did in fact take advantage of the subsidy and made the necessary modifications to become longliners. Unfortunately, by the

FV *Promise*, sibling to the *C Ola J.* (FISHERIES COUNCIL OF CANADA)

beginning of the 1960s most of the schooners that had been converted were removed from service. Age and maintenance costs coupled with their inability to raise crews signalled their demise.

Regardless of their intended role at launch, the dragger-type vessels still retained their original functional design and, with the exception of the converted schooners, were not originally developed from a Nova Scotia model. There is no question that the majority of those rigged for longlining were successful; however, most reverted to dragging before they were removed from service. What has been forgotten in Nova Scotia's history is that of all the longliner types that met the requirements of the federal Departments of Transport and Fisheries and the Nova Scotia Department of Trade, Industry, and Commerce, the Cape Island–type longliner is the only vessel that holds the official design classification by both the federal and provincial governments as being a "longliner" by design and function.

CHAPTER 13

THE BUILDUP

Yards Busy Turning Out
Nova Scotia Longliners

– HEADLINE FROM JULY 1958 ISSUE
OF *THE CANADIAN FISHERMAN*

BY 1950, THE POLITICAL CONTROVERSY BETWEEN THE FEDERAL AND provincial governments over the construction/design standards and the implementation of the Federal Fishing Vessel Construction Assistance Program as it related to individual fishers had finally been resolved. The acceptance by the federal government of a provincial design standard for the Cape Island–type longliner and the availability of funds by way of loans and the federal subsidy to the individual Nova Scotia fisher finally opened the way for the construction of the wooden Cape Island class of longliner. Similar activity was occurring with those looking for smaller fishing craft. Although there was no subsidy available for these craft, the availability of funds via loans from the Nova Scotia Fishermen's Loan Board was having an ever-increasing positive impact on the shore fishery as larger and more modern fishing craft were steadily appearing in every port in Nova Scotia and other parts of the Maritimes. With the arrival of the first longliners, especially the Cape Island type, the capabilities of the vessels in the shore fishery were starting to show.

Although the historical record confirms that the *David Pauline* was not the first Cape Island–type longliner constructed in Nova Scotia, its launch nonetheless played a pivotal role in their construction for the next twenty-five years. From a historical perspective, the *David Pauline*

Two longliners under construction, Smith & Rhuland yards, Lunenburg, NS, 1966. (NS ARCHIVES)

was the first of this class of fishing vessel to be built to a recorded federally approved provincial design standard and from formally approved drawings. From a pure technical perspective, the provincial design specification was not part of the inspection requirements enforced by the Dominion Board of Steamship Inspection but was a requirement of the federal Department of Fisheries as part of the political qualification criteria for the Federal Fishing Vessel Construction Assistance Program. Therefore, by default, the *David Pauline* was the first Cape Island–type longliner to qualify for the federal subsidy. This did not happen by chance but came as a result of the direct involvement of the Nova Scotia Department of Trade, Industry, and Commerce, which sponsored and used the construction of this vessel as a test case for the approval of the provincial design specification so needed by any future Cape Island–type longliners for qualification to receive the federal subsidy. Its construction also served to demonstrate to the federal government, especially the Department of Fisheries, that Nova Scotia did possess the engineering expertise needed to meet and exceed the requirements of the federal Department of Transport and that the risk associated with granting the subsidy to individual fishers for a properly designed Cape Island–type longliner was no different than with any other class of fishing vessel built for medium-sized fishing enterprises. Why this became such a contentious issue in the preceding four years remains a mystery.

Regardless of the political controversy that surrounded the original design of the Cape Island–type longliner, the role played by Mr. William H. Hines cannot be overstated. It is a recorded fact that Mr. Hines bore the brunt of the criticism levelled at the provincial government by the shore fishers for the government's apparent failure to recognize and support their needs. During the period that followed the Second World War, the shore fishers watched as newer and larger vessels were being built for fish processors using provincial and federal funding while they were not able to individually qualify for similar programs. Some of the criticism was well founded, especially when the Nova Scotia government publicly endorsed construction of small and medium-sized trawlers and draggers during a period when trawler restrictions were still in effect. Some government officials went so far as to propose to the legislature that the Federal Fishing Vessel Construction Assistance Program in Nova Scotia be applicable only to the construction of trawlers, draggers, and seiners. Fortunately for the shore fishery, individuals like Mr. William Hines and the local and national press of the time muted those voices. It was Mr. Hines's persistence and dogged determination between 1947 and 1949 that forced the federal government to concede to an amendment to the subsidy legislation that allowed for individual fishers to be part of the program and reduced the vessel size requirements for subsidy qualification. In addition, he can be credited with fast-tracking federal approval for the drawings for the Cape Island–type longliner and other classes of fishing vessels. For this, Nova Scotia owes him a debt of gratitude.

One question is unanswered about the government approved longliner Mr. Hines worked to develop: Was the *David Pauline* developed and built with the notion that the design would serve as a standardized pattern for the Cape Island–type longliner and as such would remain under the control of the provincial government rather than individual builders? After the vessel's launch, the drawings for the *David Pauline* were available to builders for a fee of $25 per hull, and two builders took advantage of this until their own designs were approved. Among the surviving builders and master shipwrights that worked on the construction and overhaul of the longliners, there are varying opinions on the notion of a standardized design; however, most believe the *David Pauline* was in fact designed after the Heisler half-model for the *Janet Louise*. Unfortunately, there is no documentation to prove or disprove this theory.

What is equally interesting is that given the hype over the *David Pauline*, during my research, the Robar half-model could not be located. In 1953 the Robar yard was destroyed by fire, and it is possible that the half-model could have been lost at that time; however, surviving members of the family do not remember ever seeing it. What is clear is that with the exception of a small number that were constructed in other yards from the *David Pauline* drawings, the longliners that were constructed subsequent to the *David Pauline* were not based upon the Robar design as was touted both in the press and numerous fishery-related periodicals of the time. Prior to his passing in 2019, master shipbuilder Phillip Snyder stated in an interview that in 1962, when his father decided to commence building the Cape Island longliner, his first vessel was built on a model provided by Clarence Heisler. It was from this approved design that Mr. Snyder developed his own design that in the coming years was most successful indeed. What is interesting about this transaction is that notwithstanding the competitive nature of the business at the time, the shipbuilders in Nova Scotia were a sort of fraternity that for the most part supported each other when the need arose. Over the life of the longliner, the individual builders in fact designed their version of the vessel from their own half-models, and the subsequent drawings were developed by approved naval architects. It is unfortunate that the builder developed *his* design from *his* half-model that in turn was used as the basis for the architectural drawings, yet once completed, credit for the design was often given to the naval architect.

Rather than the vessel proper, perhaps the most significant benefit arising from the launching of the *David Pauline* was the media coverage it received. In a time when communication was limited to the printed word and radio, it was media coverage that brought the Cape Island–type longliner to the forefront and notified fishers that finally the vessel they were looking for was available and more important, qualified for the

Part of the Port Bickerton, NS, longliner fleet. Notice that every vessel came from a different builder and not from a common Robar design. (WARD BOUTILIER)

Federal Fishing Vessel Construction Assistance Program. The work done by Mr. Hines made access to the critical subsidy funding possible, and the media attention paid to the launch of the *David Pauline* brought the Cape Island–type longliner to the forefront in the eyes of the shore fisher and the public at large.

Despite the media hype and the measures that were being taken to accelerate the expansion of the shore fishery, production of Cape Island–style longliners did not increase quickly. One of the major reasons for this was the requirement that builders have formal drawings developed by a naval architect and forwarded to the Department of Transport for approval in order to take advantage of the Federal Fishing Vessel Construction Assistance Program, and it took builders some time to satisfy this requirement.[80] It is therefore not surprising that the first longliners built without using the approved Robar design did not occur until March 1951.

Another issue that affected longliner construction was the limited capacity at the various yards around the province. As early as 1945, there was both considerable interest in and provincial support for the construction of small and medium-sized draggers. Construction of the Cape Island–type vessels was simply not given the same priority. Although there was some private ownership, the majority of the draggers were ordered by established fish dealers, processors, and private investors and financed by the Nova Scotia Fishermen's Loan Board and the Federal Fishing Vessel Construction Assistance Program. As explained earlier, the basic design evolved from an already known and working vessel type, and as a result, specifications had already been given federal government approval. With the plans having formal approval, the orders for these fishing craft started to increase substantially. In a number of cases, processors would order multiple hulls, thus resulting in a delay for the fishers wanting Cape Island–type longliners.

A third factor was that beginning in the mid-1940s and continuing through the 1950s, the number of smaller open vessels in the range of forty to forty-five feet was increasing, as was the construction of smaller lobster craft. A detailed review of a "builder's log" from a well-known Nova Scotia builder resulted in some startling revelations. From 1941 until 1945, a number of yards were building medium-sized harbour support craft for the Royal Canadian Navy and Royal Navy. Coupled with this, there was a steady increase in the demand for small to medium-sized fishing craft between twenty-eight and forty feet that continued after 1946. In this particular yard, evidence shows that by 1947, the yard was reaching its capacity. With the introduction of the Cape Island–type longliner, each order had to be fitted into a very tightly controlled building schedule. Most orders came on a first-come, first-served basis. By 1954, wait times exceeding eighteen months were not considered unreasonable.

Coincidental with this effort was the need for some of the shipyards to examine their infrastructure and switch from the construction of schooners and commercial yachts to the construction of these new vessels. At this particular point in time, there was a good number of qualified shipwrights available in the province to undertake the actual construction, but to most builders the longliner represented a new design with some unique requirements that necessitated the lofting of frames and other structural members from the approved drawing, and this took some time to complete. There were also issues with infrastructure. A number of the medium-sized yards had the capacity to build the traditional and now popular Cape Island snapper boat, but to construct a fifty-eight-foot vessel was just beyond the infrastructure capability of their yard. As the demand increased, a number of smaller yards like the Dover Boat Works, Little Dover, NS, increased their capability and built some very fine forty-eight-foot longliners, as did the Clarence Heisler yard at Youngs Island, NS. With the rapid buildup of the shipbuilding activity came an increased need for suitable lumber, fittings, fasteners, engines, etc. It took the lumber mills and suppliers some time to expand their infrastructure and hire the required personnel to satisfy the demands of the shipbuilding industry.[81] The logistic and material support to the shipyards was an iterative process, and most industries were able to expand their specific operations to keep up with the increasing pace of vessel construction.

In 1951, Clarence Heisler was finally able to receive "formal" approval for his design and launched the first of many longliners to follow. In the same time frame, the Harley S. Cox and Sons Ltd. and the Kenneth McAlpine & Sons Yard, both located at Shelburne, NS, came on stream and, like the Heisler yard, would play a prominent role in the history of the shore fishery and the Cape Island–type longliner. In April 1952, John MacLean & Sons Ltd., Mahone Bay, NS, known for their exceptional schooner and dragger designs, launched their first Cape Island–type longliner, as did Oscar W. d'Entremont, Lower West Pubnico, NS. By 1953 the number of builders that had approved designs and drawings for the Cape Island–type longliner was steadily increasing. Some yards such as the A. F. Thériault and Sons Ltd. facility at Meteghan River, NS, and Smith & Rhuland Ltd., Lunenburg, NS, chose to concentrate on the construction of the small and medium-sized draggers, and although both facilities designed and launched a number of very fine Cape Island–type longliners, the number launched was very small in comparison to that of the draggers and seiners launched.

By the mid-1950s the number of known yards that were building the Cape Island–type longliner was steadily increasing and ranged from the western side of Cape Breton Island with the Fraser and Chiasson yard in Chéticamp to the western side of the Bay of Fundy with the Wagstaff & Hatfield Ltd. yard in Port Greville. By 1958, the construction of the longliner was no longer confined to these yards but was taking place in small individual yards located in a number of communities along the whole coast of Nova Scotia and Cape Breton.

A fine example of the intense activity that was present in the shipyards during these busy times can be found in an excerpt from the biography of master builder George Edward Wagstaff. Mr. Wagstaff was a well-known shipbuilder throughout Canada and the United States and was one of the principal partners in the Wagstaff & Hatfield Ltd. shipyard located at Port Greville, NS. This firm was well known for its small and medium-sized draggers and seiners, and in 1956 they commenced building the Cape Island–type longliner. An excerpt from Mr. Wagstaff's personal diary, reprinted as

Appendix D, follows the construction of the Cape Island-type longliner, the FV *Danny B*, built for Captain Arthur Beatty, Port Bickerton, NS. The information in this document demonstrates the amount of coordination and workflow that had to be followed to ensure a timely completion date.[82] Note that the builder refers to the vessel by a hull number rather than a proposed name or by owner and records the date that representatives from the federal Department of Transport arrived and measured the vessel for registration. It is at this juncture that the vessel would be officially named. What is also interesting is the record of the visits by the owner and the federal and provincial inspectors for the closing inspection. This information has considerable historical value as it indicates the time required to construct a Cape Island-type longliner and confirms the level of oversight by the federal and provincial authorities. In this instance, the construction time was three and a half months. Using this as a baseline, one can readily see that the output from an average boatyard would be no more than four longliners a year. Some of the larger yards were able to produce six, but again it demonstrates the time frame required to build up the longliner fleet.

What is quite surprising is that despite the media hype and coverage that followed the launch of the *David Pauline*, only one other longliner, the FV *Clarence and Walter*, was launched from the Warren Robar Yard at Upper LaHave, NS, and actively engaged in the Nova Scotia shore fishery. Two other longliner-type vessels that followed the design of the *David Pauline* were constructed for the federal Department of Fisheries and operated out of New Brunswick and Quebec as research and support vessels. The Robar yard was rebuilt after the 1953 fire and continued to build other types of craft for the federal government and other sources but never launched another longliner.

In an effort to take advantage of the market potential presented by the construction of longliners, Harley Cox and Sons Ltd., Shelburne, NS, and Smith & Rhuland Ltd., Lunenburg, NS, purchased the rights to build Robar-design longliners while awaiting approval for their own designs. Between 1951 and 1953 a total of ten vessels were built using the Robar design, but only eight were active fishing vessels. Two were constructed at the Warren Robar Yard, five were constructed by Harley Cox and Sons Ltd., and three additional vessels were constructed at the Smith & Rhuland facility.

Construction of the Cape Island-type longliner peaked between 1955 and 1959.[83] During this five-year period, 85 vessels were brought into service, representing 41 percent of the 207 Cape Island-type longliners constructed in total during the years cover by my study (1948 to 1985). Analysis of the construction data reveals that fifty-two of the longliners built between 1955 and 1959 were fifty-seven feet or more in overall length, and seven of these were between sixty and sixty-five feet.[84] The remaining vessels ranged in length from forty-five to fifty-six feet, with the average and most popular length being forty-eight feet. By the beginning of the 1960s, yards were operating at capacity, and it was not uncommon for orders to be backed up for twenty-four months or more. During the late 1960s and into the mid-1970s there was a levelling off of longliner construction.

Between 1960 and 1985, the largest number of longliners constructed in any single year was eight in 1973. The reasons for this downward trend in vessel construction are many and varied; however, two prominent issues stand out among the others: the time required to have an order for a longliner satisfied, and changes in the fishery itself. Commencing in 1960, the number of yards that were willing to undertake the construction of a longliner was becoming limited. The demand for wooden draggers for the groundfish and scallop industries and for herring seiners for the New Brunswick fishery as well as other smaller fishing craft had increased significantly and was continuing to grow. At the same time, groundfish catches on the grounds used by the longliner were starting to decline. The major yards found that the construction of the dragger was a more profitable enterprise than longliner construction, and as a result orders for the longliner were only taken if a window of opportunity presented itself in the yard schedule. This limited the available space at the medium and smaller shipyards.

The availability of financial support for the construction of newer vessels was not limited to the shore/middle bank and offshore fishery. The aggressive marketing of the UMF resulted in the lobster industry experiencing an unprecedented period of growth and some prosperity. This prosperity in turn resulted in a number of the small ports along the coast of Cape Breton and the Eastern Shore region of Nova Scotia undertaking an overhaul and expansion of their fishing fleets; consequently, an unprecedented number of new and more efficient lobster and close-to-shore craft were being constructed. Some of the smaller yards that had undertaken the construction of longliners in the 1954 to 1958 time frame were also engaged in building the conventional snapper boat and smaller craft to satisfy the needs of the lobster and coastal groundfish industry. Although the bulk of the smaller vessels built for this element of the Nova Scotia fishery were to be used on a seasonal basis, funding was available and the construction of these vessels was seen as a very profitable enterprise for the smaller yards. As a result, there was an increasing amount of pressure being placed upon those yards that concentrated on the construction of the longliner, and orders for vessels backed up.

The expansion of the Maritime fishery brought with it the largest period of vessel construction seen in Nova Scotia since Confederation. Between 1952 and 1980, most of the major shipyards in the province were operating at capacity, and the number of yards that concentrated solely on the construction of fishing craft increased substantially. Fishing vessel construction became a major part of the economy of Nova Scotia and when coupled with the fishing industry itself, brought a measure of prosperity and growth to a number of small coastal communities. In 1966 the provincial Department of Fisheries recorded that there were forty-three yards that were approved by the Nova Scotia Fishermen's Loan Board for the construction of fishing craft ranging from large steel stern and side trawlers to small and medium-sized craft of twenty-eight feet and more. In 1974 the number of such yards increased to eighty-two. The increased number of medium- and large-sized wooden draggers being constructed for the ground and scallop fisheries, the majority of which were sponsored by the medium- and large-sized fish processors—including plants operated by the co-operative movement and UMF—required qualified fishing captains and crews. Therefore, it is not surprising that by 1957 a number of fishers who had originally expressed an interest in the longliner found themselves accepting positions as the skippers, mates, or engineers on the new draggers. Such was the case in ports like North Sydney, Arichat, Canso, Lunenburg, and Yarmouth, just to name a few.

The Cape Island-type longliner, although built in smaller numbers than other vessel types, brought a period of economic prosperity and growth to those medium and smaller yards that did not have the capacity to build draggers. And the longliner was a valuable asset to the small to medium-sized ports that had one or more medium-sized processing plants.

CHAPTER 14

MORE CHANGE

Progress is impossible without change, and those who cannot change their minds cannot change anything.

– GEORGE BERNARD SHAW

LIKE THE FISHERY IT WAS DESIGNED TO SERVE, THE LONGLINER continue to evolve and change. When the first longliners started to come off the ways in 1950, the basic design followed closely the profile of the forty-five-foot and larger Cape Island snapper boat. Like that of the snapper boat, the crew accommodation was located forward of the wheelhouse. To provide adequate headroom in this compartment, a trunk cabin ranging from eight to twelve inches in height was constructed forward of the wheelhouse. The trunk cabin contained a number of small windows to allow natural light to enter the space. Some were fitted with a small ventilator that would allow natural air to flow into the cabin area but could be closed by a watertight cover accessible from inside the cabin.

When both levels of government approved the design of the Cape Island–type longliner, it was believed that this type of fishing craft would operate on the banks and fishing grounds within fifty nautical miles from land. At the time, both the federal and provincial governments referred to this as the "near-to-shore fishery." However, by 1952, the longliners of this design were operating beyond that range. The truth of the matter was that even before the formal introduction of the Cape Island–type longliner, fishing craft of the Cape Island design were engaged in longlining on fishing grounds some seventy-five to one hundred nautical miles from the Nova Scotia coast during the late spring and summer. This was not a new phenomenon, as the impact that foreign steam trawlers and draggers were having on fish stocks in the coastal waters off Nova Scotia was forcing the shore fishers to seek and operate on fishing grounds farther from the tradition coastal grounds. This element of the shore fishery was one of the driving forces in the development of the larger fifty-eight-foot Cape Island–type longliner. The major difference was that the larger longliners, being more robust in design, was capable of remaining at sea for longer periods of time, held more fish, provided a better margin of safety, and allowed the shore fishers to continue fishing into the fall and winter.

By the end of July 1952, Harley Cox and Sons Ltd., Shelburne, NS, had completed the construction of five Cape Island–style longliners that were based on the Robar model and were sibling ships to the FV *David Pauline*. Each of these vessels was designed with the trunk cabin forward of the wheelhouse. Shortly after the vessels' launch, the owners came back to the Harley Cox shipyard and expressed concern over the potential safety hazard the trunk cabin presented. Their concerns were twofold. There was a concern that in any heavy sea, the windows in the trunk cabin might give way, thus presenting the potential for the forward cabin area to quickly become flooded. The second was directly related to the first but concerned the potential for ice to build up on the trunk cabin during winter operations. This concern was already being experienced with some of the forty-five-foot snapper boats. Research into this matter revealed that similar concerns were being expressed by a number of fishers along the Eastern Shore. Mr. Cox and his two sons reviewed the design and provided a solution that became the standard configuration for the longliners that followed.

The solution was the removal of the trunk cabin. To retain sufficient headroom in the crew accommodation space, the forward part of the longliner was raised eight inches following the forward sheer from the bow to the break deck at the forward part of the wheelhouse. This then made the forward part of the longliner all but watertight. To allow for some light to enter the forward cabin and provide a more practical exit point in the case of an emergency, a small hatch-like skylight was fitted to the forward part of the vessel approximately halfway between the bow and the wheelhouse. On each side of this structure, two small windows were fitted, made from one-quarter-inch plate glass. This design change proved to be very successful, and three additional vessels built on the original design were brought to the yard and had the trunk cabin removed and this modification embedded. It is worthy to note that a similar structure had been installed on a few smaller fishing craft built around 1950; however, the structure of the time was not as robust as the ones installed at the Harley Cox and Sons Ltd. facility in Shelburne. By 1953 this became a standard feature on all Cape Island–type longliners. By the mid-1950s this feature was being adapted by an increasing number of smaller conventional fishing craft.

In the longliner specifications developed by Mr. Hines, the structure of the gurdy house located aft of the wheelhouse was such that the port and starboard sides of the structure could be removed to allow the space to be opened in fair weather. This feature was seldom used as all of the work was carried out on the starboard side of the vessel, and most fishers believed that having the portside fixed rather than removable offered a better measure of structural safety. Therefore, it is not surprising that after 1955, unless specified by the owner, longliners were being constructed such that only the aft and starboard sides of the gurdy house could be opened. The manner in which this was done depended again upon the builder. Some builders allowed the sides to be hinged such that they could be opened and fastened to the roof of the gurdy house, while others preferred the overlapping door configuration such that the sides could be either lifted or slid out from tracks on the washboard of the gurdy house space. In either case, in inclement weather or for purposes of securing the space while in harbour, the design was such that these doors could be adequately secured and reinforced from the inside. It is somewhat ironic that these changes, however subtle, did not raise any concern with the Dominion Board of Steamboat Inspection. There is no evidence to indicate what the thoughts of the inspectors were at the time, but there's good reason to believe that the ability to open both sides of the gurdy house was seen as a convenience and that permanently fixing the portside allowed for a more sturdy and robust structure that did not compromise the original structural specification. The original specification was never amended to reflect this change.

As early as 1953, the appearance of the Cape Island–style longliner was starting to change. Fishers quickly realized that if they were to remain longer on the fishing grounds, they would need larger fuel tanks and a larger hold to accommodate larger catches and, of course, to carry the ice needed to preserve their catch. As a result of these requirements, the freeboard of the vessels started to increase. Fishers were also coming to the builder requesting that the vessels be built with more headroom in the fish hold. As a result, the vessels started becoming deeper and bulkier in their appearance. With all due credit to the builders, these changes were incorporated in all their various designs, with the end result being some of the most beautiful fishing vessels ever seen on the East Coast of North America.

The evolution continued throughout the remainder of the 1950s. Most of the modifications and/or design changes were very subtle and, in most cases, did little to change the profile of the vessel. In the majority of cases, the length of the longliner remained fairly standard. Most constructed from 1950 to 1959 were either of the fifty-eight-foot or forty-eight-foot design, with the majority being between fifty-five and fifty-eight feet. Although billed as a longliner for the groundfishery, the majority of vessels constructed during this period also engaged in the swordfishery. As a result, some of the vessels were being built with a wider stern to better accommodate two thirteen-foot double dories.[85] Also, during this period it was not uncommon to see small toilets being installed for crew comfort. As engines became larger with more horsepower, changes were made to the engine compartment and after sections to accommodate them.

In the early 1960s, some dramatic changes in the longliner design and profile started to appear. Although these changes were never universally incorporated by all builders, they did demonstrate that the longliner itself could be changed without affecting the original design standards. A review of the dimensional data contained in the official drawings for these vessels showed that in most cases the width of the vessel was about one-third the length. As an example, a vessel having a registered length of fifty-eight feet would have a breadth of approximately seventeen feet. This breadth was measured at the widest point in the vessel, which was normally located halfway between the bow and the stern. In 1962 a vessel was launched in Lower L'Ardoise, NS, that broke away from this traditional convention. The vessel was the FV *Merlin IV*. It had a registered length of 47.5 feet and a breadth of 16 feet. Unlike the conventional longliners, the breadth was carried forward to almost the forward part of the wheelhouse before it transitioned to the bow. In addition, the vessel had a depth of eight feet that matched the depth of most fifty-eight-foot longliners being built at the time. The end result was that the *Merlin IV* had a much fuller bow and higher freeboard than other longliners of a comparable length. When it first entered the fishery, some of the fishers expressed concern that the high freeboard and larger bow might cause the *Merlin IV* to pound into the sea and have a greater roll component compared to its counterparts. Surprisingly enough, this did not seem to pose a problem, and it wasn't long before other vessels with similar dimensions started to appear in the builders' yards. It is somewhat ironic that this vessel and others like it opened an era in fishing vessel construction where both the medium and small-sized vessels were being launched with a much larger profile than was thought possible. With the introduction of fibreglass and composite building materials, this trend continues to the present day and has allowed for the construction of coastal fishing craft—many of which are based on the traditional Cape Islander—where the width is almost half of the registered length.

In the mid-1970s and continuing on to the introduction of fibreglass and composite materials, there was another subtle change in longliner design. The introduction of the wider and fuller bow resulted in more space for accommodation forward. This in turn resulted in the wheelhouse moving farther forward, allowing more working space aft. Some builders replaced the wire guardrails with solid bulwarks. None of these changes were seen as a compromise to the original design specification but rather an improvement to structural integrity and crew safety and comfort.

In the original design documents, there was always the convention of having the face of the wheelhouse slope slightly aft. There was a belief that this slope allowed seas to break over the wheelhouse much more effectively, thus reducing the stress on the structure. There is no documentation to show that there were any formal studies conducted to support this theory, and there is nothing in the government approved specifications that refers to this design feature. It was common knowledge that vessels of similar size working in the West Coast fishery had as part of their design the face of the wheelhouse sloping forward. The reason given for this configuration was that it provided better forward visibility and reduced the internal glare from the wheelhouse when sailing in darkness. Whatever the reason, this design feature started to be incorporated in some of the longliners constructed in Nova Scotia from the late 1960s onward. It is also worthy to note that most of the longliners that underwent major refits after 1970 and had their wheelhouse replaced incorporated the forward sloping face on the wheelhouse. This wheelhouse configuration proved quite successful and is now standard on most small and medium-sized fishing craft being constructed in Nova Scotia building yards today.

CHAPTER 15

THE NEWFOUNDLAND CONNECTION

*It may be that in the near future the plant
with some government assistance as at
present can absorb a large share of the
capital cost of the longliner and if so a
profitable operation for both plants and
fisherman is possible over much of the area
of North East coast of Newfoundland.*

– FISHERIES RESEARCH BOARD OF CANADA, 1963

IN THE LATE 1930S AND CONTINUING INTO THE 1940S, THE PROBLEMS that were being experienced by the Nova Scotia shore fisher did not go unnoticed by the then British colony of Newfoundland. Newfoundland had entered the twentieth century with a very lucrative and success- ful fishery. Its proximity to some of the richest fishing grounds in the world allowed it to become very competitive in the salt fish market, and in many cases its production and sales outstripped that of Nova Scotia. Despite its early successes, the Newfoundland fishing indus- try was rapidly becoming outdated. By the late 1930s, sailing vessels were giving way to motor craft, and the small inshore fishers were finding it increasingly difficult to compete. As in Nova Scotia, foreign steam trawlers were fishing the traditional coastal fishing grounds, especially of the south and eastern coasts.

By the time Newfoundland entered Confederation in 1949, the inshore fishing fleet was made up of about 19,500 fishers, all small-boat operators and dispersed in a host of small, isolated villages scattered along 9,600 kilometres of rugged coastline. The fishers toiled unceasingly from small boats using antiquated cod gear and methods only to receive less than $500 annually for their efforts. Prior to entering Confederation, the officials responsible for the fishery in the Newfoundland colony watched with interest the slow and steady growth of the Cape Island fishing vessel and how the Nova Scotia shore fishers were actually longlining from these craft rather than using the traditional dory. Shortly after joining Confederation, the newly formed provincial government of Newfoundland developed a plan to modernize the Newfoundland fishery. As a first step, the provincial officials believed it important to investigate better vessels, fishing methods, and marketing strategies and then scope out how these elements could be combined to form a comprehensive and viable plan for the provincial fishery. One of the first priorities of this plan was an attempt to replace the small, open fishing craft used by the inshore fishers with a modern, moderate-sized powered craft that could operate longer during the year. Of prime interest to Newfoundland was the Cape Island–style snapper boat and the ongoing development of the Cape Island–type longliner.

Having identified a potential fishing platform, the government of Newfoundland in conjunction with the federal Department of Fisheries expanded the investigative process further to examine all facets of how longlining could be profitably employed in the near and middle bank fishery around Newfoundland waters. The decision was made to conduct a series of precise experiments that would look at the suitability of vessels, the type and quantity of catch, a comparative analysis of the catch between longliners and schooners, the geographical location, depth of water, scientific and biological data such as water temperature, seasonal variation of bait, and the limitations that weather would place on the longliner from a geographic and seasonal perspective. Coincidental with this initiative was a further plan that would examine the commercial viability of a series of geographic areas relative to the number of fishing ports and identify the types and locations of processing facilities that would be required to make the fishery profitable.

After considerable discussion, and with the support of the federal Department of Fisheries, officials from the government of Newfoundland came to the South Shore of Nova Scotia in the spring of 1950 and chartered two Cape Island snapper boats: the FVs *O-Johnny-O* and *Robert and Donald*, both based in their home port of Lockeport, NS. These vessels remained in Newfoundland under charter from June through December 1950. The following spring the vessels remained under charter and were joined by another vessel, the FV *Miss Osborne*.[86] While the experiments revealed some very positive results, they clearly demonstrated that there were areas where longlining, although viable, might not be profitable in the mid and long term.

The experiments continued in the spring of 1952, and this time four Cape Island-type longliners were chartered. Three of the vessels were based on the Robar design: the FVs *Pat and Judy*, *Marjorie Byrl*, and *Gertrude and Ronald* built by Harley Cox and Sons Ltd., Shelburne, NS. The fourth vessel, the FV *Atlantic Rover*, was designed and built by Kenneth McAlpine and Sons Ltd., also of Shelburne. The experiments were completed in the fall of 1955. At that time, the vessels returned home to Nova Scotia and engaged in the commercial groundfishery.

The extensive research carried out by the Province of Newfoundland and the federal Department of Fisheries between 1950 and 1955 marked the first time that any governmental agencies followed a structured approach to evaluating new fishing vessel design, fishing methods, and their application to the entire geographic area of a shore fishery. The depth of the trials clearly demonstrated that the forty-to-forty-five-foot Cape Island–style fishing craft was an ideal platform for longlining operations in the near-shore fishery along the coast of Newfoundland and Labrador, and this capability was further enhanced with the introduction of the larger Cape Island–type longliner. The trials team recognized that the Cape Island–type longliner was not the only fishing platform suitable to the Newfoundland fishery. There was unanimous agreement that the use of large offshore trawlers and smaller draggers with their capability for larger catches would definitely play a major role in the expansion of the fishery. However, the shore fishery central to the hundreds of small and isolated communities along the coast of Newfoundland and Labrador was considered vital to the survival of the majority of the residents of these areas, and much like the situation in Nova Scotia, the shore fishers needed better boats and more effective fishing methods. In these circumstances it was concluded that the introduction of the large and medium-sized trawlers was not practical. Again, there was unanimous agreement that the ideal platform for the shore fishery was the Cape Island–style fishing vessel, with the preference being given to the new Cape Island–type longliner.

It remains unclear as to what effect the results of the research program had on the Nova Scotia shore fishery or the future of the Cape Island–type longliner. It does make one question why a similar research program was not conducted in Nova Scotia in the 1945 time frame when the popularity of the Cape Island–design fishing craft was steadily on the increase, instead of attempting to identify a common hull form and do so without a structured approach to the problem. In this regard Newfoundland was well out in front of Nova Scotia in its holistic approach to the expansion of the fishery.

Despite the success and thoroughness of the research program, the Province of Newfoundland faced a major problem. In 1955, the province did not have access to the drawings for the Cape Island–type fishing craft and, given that the models and plans were the property of the individual builders, the Province of Nova Scotia could not intercede or provide any drawings to Newfoundland. Although a number of shipyards were operating in Newfoundland at the time, most of their expertise was in the construction of schooners and other medium-sized sailing craft. Some of the small yards were building the familiar motor dory–type fishing craft for the smaller ports along the coast, but none possessed the expertise to undertake the construction of the Cape Island–style fishing vessel, especially the Cape Island–type longliner. Initially, Newfoundland officials thought this problem could be resolved in the short and medium term by having the longliners built in Nova Scotia and at the same time inviting builders to come to Newfoundland and assist the builders there with establishing the construction of the vessels in that province. This notion failed to materialize, as by 1955 the majority of the yards in Nova Scotia were at capacity constructing longliners and/or small draggers. In some instances, the orders were booked one year ahead. Therefore, it could take five years or more before longliners would start showing up in Newfoundland ports. Through the Province of Nova Scotia, the Newfoundland government approached a number of builders and offered financial assistance for them to come to the province and establish satellite shipyards and construct the Cape Island–type longliner.[87] Unfortunately for Newfoundland, the activity and building boom in the Nova Scotia yards was such that no one was interested in relocating.

The second problem that faced provincial authorities in Newfoundland was that the Federal Fishing Vessel Construction Assistance Program applied only to new construction or the conversion of schooners to longliners or draggers. Given that most of the schooners were obsolete and not cost effective to convert, the only option was new construction. Unless a solution could be found that would allow new vessels to be constructed either in Newfoundland or Nova Scotia, the fishers would not qualify for federal assistance. The Province of Newfoundland was able to somewhat mitigate this problem by providing assistance to those fishers wishing to buy Cape Island–type snapper boats that came available when their owners decided to upgrade to the Cape Island–type longliner. Between 1955 and 1960, a number of the Cape Island snapper boats in Nova Scotia were purchased by Newfoundland fishers and transferred to Newfoundland. By the mid-1960s, Newfoundland was able to develop a design and get approval for the construction of a Cape Island–type longliner. It was from this design that medium-sized fishing craft started to appear in increasing numbers along the Newfoundland coast. In this same time frame, a number of Nova Scotia longliners became available for sale, and the majority were purchased by Newfoundland interests.

It is interesting to note that the Newfoundland experiments marked the first time that any serious attempt was made to gather data on the size and health of the cod and other groundfish stocks on the inshore fishing grounds surrounding Newfoundland and the coast of Labrador. Unbeknownst to the scientists of the day, the data they gathered would be used in the mid-1980s as the baseline for determining the effects of overfishing on the fish stocks and the subsequent decision to close the cod fishery in the early 1990s.

CHAPTER 16

LONGLINER CONTROVERSY

The loss record of fishing boats in the coastal and offshore waters of Nova Scotia shows that the Cape Island model compares favourably with the models of recognized naval architects and other boat designers.

– W. S. HINES, NOVA SCOTIA DEPARTMENT OF TRADE, INDUSTRY, AND COMMERCE, 1954

IN 1953 THE FISHERIES DIVISION OF THE FOOD AND AGRICULTURE Organization (FAO) of the United Nations, headquartered in Rome, Italy, started to explore the notion that the world's oceans provided a renewable food resource that hitherto had been neglected and held the possibility for greater development. The FAO also believed that if the sea's resources were to be properly exploited and developed, then it was equally important to pay greater attention to the construction of fishing vessels worldwide.

As a starting point for discussions on this important issue, in late 1952 the FAO commenced the planning for a major international congress that would provide a forum to discuss the designs of present and future fishing craft of all types and rigs. The intent was to open the attendance to naval architects, engineers, and fishers from all over the world. The founders of the congress believed that having such a gathering of world-renowned architects and engineers in a single body would result in improvements to the design and engineering of future fishing vessels.

In 1953 the FAO organized two separate congress meetings. The first meeting was held in Paris in April and was mainly to solidify the format and content for future congress meetings. The first congress took place in November 1953 during the Gulf and Caribbean Fisheries Institute Convention in Miami Beach, Florida. The congress attracted over two hundred fishing craft designers, engineers, architects, scientists, and fisheries representatives from all over the world. One of the Canadian contributors to the conference was none other than Mr. William S. Hines, Chief Engineer, Nova Scotia Department of Trade, Industry, and Commerce, Halifax, NS.

One of the American contributors to the congress was noted American architect, designer, and author Howard I. Chapelle. Although a very well-known and respected architect, Mr. Chapelle's area of expertise was confined to the design and survey of ship and boatbuilding programs for the United States Army transportation system during the Second World War and the research of a number of colonial-designed sailing craft for European interests after the end of hostilities. Other than his engineering knowledge of structure and stability, at this point in his career Mr. Chapelle had very limited exposure to the design of fishing craft.

At the 1953 congress, Mr. Chapelle presented a paper entitled "Some American Fishing Launches." The paper covered the design parameters of a number of small fishing craft types including the Gaspé boat, the American Seabright skiff, and others, but most notable among them was his scathing attack on the Cape Island–design fishing boat and its cousin, the Jonesport design. In his presentation Mr. Chapelle went to great lengths to state that for its size the Cape Island boat was extremely light, was constructed of material that was thin in the cross-sectional area, and had a serious design flaw in that the stern post and keel were merely bolted together at the keel, as was the horn timber and stern. His presentation continued to attack the design, saying that the long sharp bow combined with a broad flat stern and shallow draft produced a hull that would broach in a heavy following sea. In the end there was not one aspect of the Cape Island design that found favour with Mr. Chapelle. In his view, the Cape Island/Jonesport design was not a safer replacement for the older style fishing craft that were based on the schooner/whaler design. Throughout his presentation, he made it very clear that a number of fatal losses along the Eastern Seaboard of the United Sates were directly related to this design and stated that as a result of the serious design deficiencies, hardly a season passed in which there were not a number of fatalities in this class of fishing vessel. In his view, a completely new design was required.

Excerpts from his paper, including his condemnation of the Cape Island design, were published in the October 1953 edition of *The Canadian Fisherman* magazine, almost one month before his paper was presented at the Miami congress. Releasing extracts from his paper prior to his formal presentation was definitely unorthodox. Why Mr. Chapelle did so remains a mystery. Some sources believe that given his limited background in fishing vessel design, the early release of the paper could have been a means for him to gauge the response to it prior to the conference. If his intent was to seek feedback prior to the conference, his criticism of the Cape Island design definitely achieved the aim. Mr. Chapelle's criticism of the Cape Island design was not well received by the Nova Scotia builders and fishers and totally outraged Mr. William S. Hines, the chief engineer from the Nova Scotia Department of Trade, Industry, and Commerce—and with good reason.

The subject Mr. Chapelle had chosen for his paper was the Jonesport boat. Although there were some similarities between this craft and the Cape Island boat, the actual structural design was totally different. In addition, the platform used by Mr. Chapelle as the basis for his argument and criticism of the Cape Island design was one of the very early designs of the Jonesport boat that represented the transition from sailing craft to a motorized vessel. This design had its beginnings somewhere in the early twentieth century, and as a motor craft, it was quite rudimentary. While there is no question that there were similarities between this vessel and the early Cape Island design, starting in 1907, the Cape Island design began to become quite distinct. While the Jonesport design remained relatively stable both in dimension and structure, the Cape Island boat became broader with a more distinct sheer. There were also some major structural differences. The Cape Island boat was built from heavier materials with the emphasis being placed on those types of wood that were known for their strength and resistance to rot. So although there were outward similarities between the Jonesport and the Cape Island designs in the beginning, the Cape Island design continued to improve. With each improvement the Cape Island moved further away from the Jonesport design. It continued an evolutionary process that took many forms and increased its popularity as a multipurpose fishing platform that by 1948 had evolved and formed the basis of the Cape Island–type longliner. The Jonesport design evolved, but not to the extent that the Cape Island boat did.

In developing his argument, Mr. Chapelle failed to recognize that there was a significant difference between the Jonesport boat and the Cape Island designs of the late 1930s and early 1940s, and he made the critical error of comparing the Cape Island design criteria and structural data with fishing vessel designs that had been obsolete for thirty years. Mr. Chapelle was one of the representatives at the first American Bureau of Shipping standard conferences of 1942. It is rather surprising, then, that in his critique of the Cape Island design in which he mentions the longliners, he not only makes no reference to the fact that the Cape Island fishing craft had undergone a number of major modifications since the original design in 1905, but also neglected to mention that the longliner was now subject to a set of national and international building standards that he had helped to develop.

Whether or not it was intentional, Mr. Chapelle's paper seeded some doubt as to the safety and structural integrity of the Cape Island design. It certainly struck a nerve with most Nova Scotia builders and was most unwelcomed by the Government of Nova Scotia. For a naval architect of Mr. Chapelle's standing to level such unfounded criticism on the Cape Island design in such a prestigious international forum was considered a slap in the face to the Nova Scotia shipbuilding industry. Mr. Chapelle's presentation laid the foundation for a controversy that would last for years to come. Within the Nova Scotia government, there was no one more upset than Mr. Hines. When the extracts from Mr. Chapelle's paper appeared in the October 1953 edition of *The Canadian Fisherman*, Mr. Hines knew full well that the data being used to develop the argument was based on obsolete sources. He also knew that Mr. Chapelle's comments would cause considerable problems within the Nova Scotia shipbuilding industry at a very critical time and there had to be a way to defuse the situation quickly without starting an international argument. Although there was ample time for Mr. Hines to read and respond to Mr. Chapelle's criticism and arguments before attending the congress, he chose not to do so. Instead, Mr. Hines attended the conference and allowed Mr. Chapelle to present his paper. At some point during the discussion, Mr. Hines openly challenged Mr. Chapelle's remarks and arguments. Mr. Hines's own remarks and arguments were so strong and exacting that Mr. Chapelle had to concede that he was not fully informed nor totally knowledgeable of the Cape Island design and the improvements that had been made since its humble beginnings more than forty years earlier. He also conceded that he was not fully aware of discussion details and standards being used to build the new Cape Island–type longliners. In the end, at Mr. Hines's bidding, Mr. Chapelle modified his statements for the record to agree that the difficulty encountered by the boats in the past was due to the use to which they were put rather than faulty structural or design characteristics of the platform.

Upon his return to Canada, Mr. Hines spent considerable time addressing the concerns raised by Mr. Chapelle. It was not until August 1954, ten months after the excerpt of Mr. Chapelle's paper was published, that *The Canadian Fisherman* published Mr. Hines's rebuttal to Mr. Chapelle's criticism. Mr. Hines should be commended for the manner in which he approached the rebuttal; it was quite obvious that he was doing his utmost to defuse the controversy and bring the issue to closure. In his article, Mr. Hines provided a brief yet fairly detailed history of the Cape Island design and how over time the Cape Island boat became one of the best loved and best known fishing platforms in the Nova Scotia fishery. He also took great pains in explaining how the Cape Island design had grown from a small motorized fishing skiff to a fifty-five to fifty-nine-foot longliner and made a point of stating the inspection criteria that applied to each vessel over fifteen tons. Mr. Hines very diplomatically stated that the member who had tabled criticism at the conference did in fact modify his comments. Not once in his rebuttal did Mr. Hines refer to Mr. Chapelle by name or make any personal reference to him. It is very ironic that Mr. Chapelle, who was so critical of the Cape Island design, was within two years completing drawings for the Cape Island–design longliner from half-models developed by the Smith & Rhuland shipyard in Lunenburg, NS, and that Mr. Chapelle's drawings, which became the signature design of the firm, had to be approved by none other than Mr. William Hines.

In October 1954, Mr. Chapelle, who by now had taken up residence in Mahone Bay, NS, and was doing some design work for John

MacLean & Sons Ltd., wrote a letter to Captain A. T. Muir, the principal editor of *The Canadian Fisherman*. In his letter, published in the October 1954 edition, Mr. Chapelle tried to refute Mr. Hines's recollection of the events that prompted his retraction and went on to try to justify what he had said. He stated that his paper did not intend to cast doubt on the design of the Cape Island–type longliner but earlier versions of the Cape Island design. Regrettably, the damage had already been done.

The controversy that resulted from Mr. Chapelle's comments did not totally disappear from the Nova Scotia shipbuilding industry in the years following the congress. There were those proponents of draggers in the hundred-foot range that used his original comments to support their argument that it was a waste of money to continue building longliners, as they were dangerous and totally unsafe for the near-shore fishery. On the other side were those in small communities who believed that the longliner was in itself a very versatile and valuable asset to the shore fishery but had to be managed properly and operated with in its design limits. In the mid-1950s there had been some loss of life involving longliners, but a comparison of the accidents involving longliners with those of other fishing fleets found that the losses involving longliners were fewer than average.

In March 1961 tragedy struck the small fishing town of Lockeport, NS. Three longliners, FVs *Marjorie Bryl*, *Jimmy and Sisters*, and *Muriel Eileen* were caught in a major storm and lost with all hands. The tightly knit town lost seventeen well-known and respected fishers, some of whom were from the same family. The tragedy brought the community to its knees with grief. The exact cause of the losses will likely never be known. The fishers had left for the fishing banks off the coast and got caught in a non-forecasted winter storm. There were a number of vessels along the coast of Nova Scotia from Lockeport through to Canso that were caught in the same storm and made it back safely to port, but unfortunately these three failed to return. On May 12, 1961, the Honourable Lloyd Crouse, member of parliament for Queens Lunenburg, the constituency that included the town of Lockeport, arose before the Committee of Supply and Fisheries in the Canadian Parliament and unleashed a scathing attack on the longliners.[88] Mr. Crouse said the Government of Canada should be very concerned with the loss of seventeen fishers. He stated in no uncertain terms that the longliners were never intended to operate over one hundred miles at sea and that there was a structural weakness in these craft that needed to be examined. He criticized the fact that the large wheelhouse placed amidships covered the engine and that the fish handling area/gurdy house had been designed as a shelter from the elements rather than being an integral part of the whole vessel and when struck by heavy sea could quickly be destroyed, leaving the vessel exposed to the elements. He then expanded upon the virtues of the sixty-five-foot dragger and ships of the schooner type that were between seventy-five and one hundred feet in length. In his address he was adamant that this unfortunate event required the committee to conduct a complete study of the Federal Fishing Vessel Construction Assistance Program and went on to state that the money and subsidies would be better spent on the larger draggers than in the construction of longliners.

Mr. Crouse's comments in the House of Commons rekindled the controversy over the design of the Cape Island longliner. His comments were covered by national press and again brought about a degree of doubt as to the structural integrity and safety of the Cape Island design. In the July 1961 issue of *The Canadian Fisherman*, the editor published an editorial that highlighted the Honourable Mr. Crouse's comments and related his comments back to those made by Mr. Chapelle in 1953. This time, surprisingly enough, there was no formal rebuttal by any official of the Nova Scotia government or representative from the Nova Scotia shipbuilding industry. It remains unclear as to why this matter was left to fade away, but there is some conjecture as to why Mr. Crouse's comments failed to go any further than in the notes and proceedings of the committee itself.

There is no question that the Honourable Lloyd Crouse was one of the most distinguished and respected members of the Canadian House of Parliament in the history of this great country. He was known for his honesty, integrity, and concern for the people he represented. If there was one failing that may have affected his approach to the longliner design, it was the fact that Mr. Crouse, prior to entering the political arena, was himself a major fish dealer who owned and operated deep-sea draggers and was a proponent of their use. In making his presentation to the committee, his appeal to have the subsidies apply to the dragger/schooner-type vessels may have been seen by many, especially in this province, as self-serving, and his argument may have been self-defeating.

There is no question that the loss of life in Lockeport did raise concerns over the safety and structural integrity of the longliners to the point that some people started referring to them as "widow-makers." Whatever the depth of concern was, there was no intervention by the provincial authorities or the federal Department of Transport save for the investigation into the loss of the three vessels themselves.

CHAPTER 17

THE LONGLINER AND THE FISHERY, 1949–1985

We used to call longlining "trawl fishing" around here. Had a couple of boats. Loved it. Best way to make a living on God's good green ocean.

– BLAIR TINKER, FISHER

WHEN THE DESIGN OF THE CAPE ISLAND–TYPE LONGLINER WAS FINALLY approved in 1949 and was included as part of the Federal Fishing Vessel Construction Assistance Program in 1950, it was done to satisfy the requirement for a medium-sized fishing craft for the shore fishery. This vessel took a long time coming, but after two decades of political wrangling it finally was able to make its presence felt. Based on the dynamics of the fishery at the time, both the federal and provincial departments responsible for the design approval process believed that the prime species of interest were cod, haddock, hake, pollock, and halibut. It was also their belief that the prime harvesting method of the shore fishery would be the use of long lines of baited hooks that would be set and retrieved from the vessel itself, a method made possible by the change from sail to power. This can be substantiated by the fact that both the federal and provincial governments established the

Cape Island–type longliner as being a class of fishing vessel. Further, the larger size of the Cape Island–type longliner meant more trawl could be carried, so the government had as one of the requirements for this new class of fishing vessel the need for the vessel to have a mechanical trawl hauler known as a "gurdy." What remains unclear is the perception both governments had with regard to the employment of this vessel. Was it their belief that this vessel would replace the smaller fishing craft of the day, or was there a notion that this vessel would complement the coastal fishing fleet, and if so, what would be the expected numbers and the projected impact? There is no evidence to show that any studies of this nature were ever undertaken. It would appear that although there was a recognized need for the longliner, its place in the Nova Scotia shore fishery remained uncertain.

The fishers' perspective was somewhat different. In the course of my research, I held multiple interviews and informal discussions with over twenty-five fishers, builders, and fish plant owners. During my conversations with twelve former longliner owners/captains and two fish plant owners, there was a constant theme that remained consistent with each individual. That theme was the important role the Cape Island–type longliner played in their lives as fishers and plant owners. The fishers I interviewed were experienced shore fishers from various fishing communities located throughout Nova Scotia who had owned or crewed medium-sized Cape Island–style fishing craft from the mid- to late 1940s prior to the purchase of a longliner. One gentleman actually started his career at the age of fifteen in the Grand Banks fishery working trawl lines in a dory from a Lunenburg schooner. The age spread among the fishers I interviewed was approximately fifteen years. Despite the time span and the changes that would have occurred in the fishery during that period, each fisher without exception believed that he did better financially by investing in and moving up to the larger Cape Island–type longliner. They also stated that they could fish longer during the year than they could with the smaller, less capable Cape Island–style snapper boat. In addition to longlining and other types of fishing that arose during the period under study, all engaged in the groundfishery and swordfishing, while a few had also engaged in Danish seining. All the fishers that had engaged in the swordfishery, including the youngest of the group, considered the swordfishery to be one of the critical factors in purchasing a longliner. In fact, as a young man, the youngest of the group began his career in a Cape Island–type longliner with his father and had another built when he had the opportunity to take command of his own vessel. With the exception of one individual, all of the fishers that engaged in summer swordfishing considered this activity to be equally or more important than longlining for groundfish or any other type of fishing. All agreed that the presence of foreign and Canadian draggers on the traditional inshore grounds had a major impact on the fish stocks and the Nova Scotia fishing industry as a whole. As a result, the fishers had to go farther offshore, thus larger vessels were required. Without exception all had engaged in longlining for groundfish until the late 1950s and early 1960s as this was the only "technology" available for harvesting the fish save for transitioning to a small dragger that at this point in time was out of their financial reach. The two gentlemen who had ownership of three medium-sized fish plants agreed that the introduction of the Cape Island–type longliner played a major role in the stability and profitability of their operation.

As I walked along the various wharves in smaller Nova Scotian communities, I had off-the-cuff conversations with a number of older fishers who are of the opinion that the longliner arrived too late. It's a belief of some that by the time the shore fisher was able to purchase the vessel he needed, the devastating effects the draggers were having on the inshore and middle banks had already taken hold, and that the longliner fishery was doomed before it ever started. Most of these conversations classify as what we in Nova Scotia would call "wharf talk." Most of these gentlemen never owned a longliner, but the vast majority of them went to sea as crew and fished on them for a number of years. An examination of this fishery reveals that there is considerable truth in what these older gentlemen are saying. However, what people have failed to recognize is that although originally conceived and designed for a specific aspect of the Nova Scotia fishery, over its thirty-plus years, the Cape Island–type longliner became involved in almost every aspect of the shore fishery. No other vessel before or since can make such a claim.

When the longliner started to appear in the fishing ports, it added a new dynamic to the groundfishery. The first Cape Island–type long-liners, like the FVs *Janet Louise*, *David Pauline*, and others, were designed, it was believed, to operate within fifty nautical miles from shore. However, by 1953 and onward the profile of the vessel changed substantially. The longliner grew both in length and gross tonnage. The vessels coming off the ways had a higher freeboard, were deeper between the aft deck and bottom of the fish hold, and had a deeper draft. By 1954 it was not uncommon for some to be operating on the fishing banks in excess of one hundred nautical miles from the coast. Most were capable of staying at sea for up to five days, though most trips lasted an average of three to four days, not including the transit to and from the fishing banks. All were capable of carrying ice to keep their bait and catch from spoiling and depending upon the geographic location of their home port were capable of fishing during the late fall and winter months. There were a number of factors—such as weather conditions and Arctic pack ice, especially on and around the grounds of the Gulf of St. Lawrence—that determined the length of any one individual season, but across the whole of the longliner fleet from 1950 to 1985, the time at sea averaged 280 days.[89] This was a significant increase from the days accumulated by the smaller Cape Island snapper boats, which depending upon the location would average at best 200 days. In locations around Cape Breton the time was much less.

By 1960 there were 104 Cape Island–type longliners operating out of ports from Chéticamp to Yarmouth. In the beginning, the longliners did quite well, and catches across the board were averaging between fifteen and twenty thousand pounds of mixed fish for a four-day trip. Despite the market conditions, the price of groundfish remained relatively low, but so was the price of bait, fuel, and other necessities that were required for the trip. The owner normally took a percentage of the catch to cover the capital cost of equipment, boat payments, and maintenance.[90] This was normally known as the "boat share." Once the expenses were deducted, the remainder was divided up evenly among the crew. On a good three-to-four-day trip, the crew could clear anywhere from $80 to $100 for their efforts, an amount that was considered exceptional for its time.[91] The collateral benefits from this catch extended to the local fish plant. The catches being landed by the longliners, coupled with those from the smaller boats, resulted in increased employment in the small to medium fish plants. This employment boosted the local economy, and the once poor villages were seeing a measure of growth and stability. Depending upon the geographic location along the coast, a good number of the longliners could continue to fish well into the fall and early winter. This allowed the local fish plants to operate longer, thus increasing local employment even further. Between 1952 and 1965, the fishing communities that had longliners operating from them enjoyed a period of unprecedented economic growth and prosperity. For these communities the future looked very promising.

By the early 1960s, the continued buildup of large domestic and foreign trawler and dragger fleets on the fishing banks adjacent to the Nova Scotia coast were having an ever-increasing negative effect on the longliner operations. The average catch per vessel fishing on the same grounds as they did prior to 1960 started to show a sharp decline. Catches were now two-thirds to one-half of what they were just four years previous, while the price of fish increased less than 5 percent. The average catch for a three-to-four-day trip was now between nine and twelve thousand pounds. This, combined with the fact that the smaller coastal boats were also experiencing smaller catches, was starting to have a negative impact on the economy of a number of small fishing communities along the Nova Scotia coast.

The fishers again complained about the incursion of domestic and foreign draggers on their traditional fishing grounds that in most cases were outside the three-mile limit on territorial waters. As an example, between 1954 and 1956 Russia alone operated fifty-two large trawlers, four mother ships, and two other support vessels on the fishing banks between Newfoundland and Nova Scotia. Each year, the number of foreign trawlers increased, as did their overall size. A number of these vessels, along with smaller draggers from other countries, were fishing on the inside edge on the Scotian Shelf between Sable Island and mainland Nova Scotia. A large number were often found operating adjacent to the three-mile limit and in some cases well within visual range from shore.

The plight of inshore fishers, including those that operated longliners, was slowly gaining some traction with the federal government. In 1964, after a host of unsuccessful negotiations at the United Nations Law of the Sea conferences, Canada unilaterally declared that its territorial waters extended to twelve nautical miles. In 1971 Canada again declared and established exclusive fishing rights for Canadians in the Bay of Fundy and Gulf of St. Lawrence.[92] Although this was seen by some as a very positive and progressive step to protect the shore fishery, for those involved in longlining for groundfish, the damage to the fish stocks had already been done, and the once rich fishing grounds off the east and south coasts of Nova Scotia were still open to exploitation by domestic and foreign draggers. Perhaps the most positive point that came from the actions taken by the government at that time was the protection of the halibut grounds in the Gulf of St. Lawrence.

The effect of foreign and Canadian seiners operating in the coastal waters off Nova Scotia also caused a sharp reduction in baitfish stocks such as mackerel, herring, and squid, thus driving the price up in some instances by 100 percent. The costs associated with normal maintenance, fuel, and other essentials were also increasing and were outpacing the price the fishers were receiving for their catch. The longliner fishers were still able to make a reasonable living, but it became very evident that the depletion of fish stocks on the traditional fishing grounds brought into question the long-term viability of this method of fishing. By the beginning of the 1960s, and again depending upon the geographic location of the home port, a good number of longliners engaged in other methods of fishing during the spring, fall, and winter months. Others could no longer fish from their home port and had to migrate to ports that were closer to more lucrative fishing grounds. As an example, longliners from Canso and Whitehead often migrated to Cape Breton coastal ports such as Louisbourg, Port Morien, or Glace Bay in the spring and late fall to be closer to the area around Scatarie Bank and the adjacent fishing grounds. In such cases they often sold their fish locally rather than returning to their home port. As a result, a number of the plants in the small Nova Scotia communities were forced to sell to larger entities or go out of business completely.

Between 1962 and 1980, the number of fishing craft of all types engaged in longlining started to decrease. The number of smaller close-to-shore boats remained relatively stable until the mid-1970s, but it was not so with the larger longliners. Some, especially those located in ports along the South Shore of Nova Scotia, remained actively engaged, but by 1976 the majority of the Cape Island–type longliners had converted to other types of fishing.

Next to the lobster fishery, swordfishing was perhaps one of the most important fisheries to the shore fishers, and it could be argued that from an economic standpoint it was more important than longlining for groundfish. By the end of the Second World War, the swordfish that had been so prominent in the coastal waters off Nova Scotia had moved to the outer fishing banks and could be found in some abundance from Georges Bank to the Flemish Cap. Although Nova Scotia longliners fished this vast expanse of sea, the majority of those hunting for swordfish operated between Georges Bank and the waters around Sable Island. The increased size of the Cape Island–type longliner allowed fishers to operate during the summertime in these waters that were an average of one hundred nautical miles from the coast. The primary method of harvesting swordfish was by visual sighting and harpoon. The price paid for the fish was tenfold what one received for groundfish, and there was no requirement to maintain expensive trawl lines. In the early 1950s, the longliners would rig up in early June and fish until late August or early September. The crew normally consisted of five, including the captain, and the trips normally lasted five days. During this three-month period, those engaged in the swordfishery did very well financially and in some instances matched or exceeded the earnings gained from the spring, fall, and winter groundfishery.

Not all of the Cape Island–type longliners went swordfishing. The distance to and from Sable Island and the surrounding banks such as Georges and Browns was considered too far for some of the longliners operating out of ports in northern and western Cape Breton, such as Chéticamp. There were a few that rigged up and went, but rather than returning to home port, they operated out of ports like Louisbourg, Port Bickerton, and Canso. A good number chose to remain engaged in longlining. Because of the distance, most of those longliners engaged in swordfishing were the larger vessels of fifty-five feet or more in length, although some smaller longliners from along the South Shore of Nova Scotia were also involved. The smaller vessels were able to become engaged mainly because of their proximity to Georges and Browns Banks. Very few ventured to the waters off Sable Island.

From the time the longliners started to be employed in the early 1950s through to the early 1960s, the swordfish stocks remained reasonably healthy, but there was some evidence that this would not last forever. Although the numbers taken remained relatively the same, the fish were smaller and the landed weight showed some decline. Given that the waters off Nova Scotia where the swordfish carried out its summer migration were at the time declared international waters, the swordfish came under increasing pressure from not only Canadian fishing craft, but also a very large American fleet operating out of ports from South Carolina to Maine.

In the late 1950s American swordfishers started to longline for swordfish using a method similar to that used for harvesting groundfish. The main trawl line was obviously of a heavier gauge than that used for groundfish, and the hook was much larger. Unlike the groundfishery, the swordfish trawl did not go to the bottom but was set at a depth where the swordfish would hunt for food. This trawl line was set with a buoy at one end, others along its length, with the last end remaining attached to the vessel. The vessel and the trawl were then allowed to drift for a set period of time after which it was retrieved. The Americans had great success with this innovation and were increasing their landings substantially.

In the summer of 1962—under the sponsorship of the Nova Scotia Department of Trade, Industry, and Commerce—the Sambro, NS, longliner *George and Pauline* under the command of Captains Louis and George Hennebury rigged up and conducted a number of experiments with respect to using trawl lines to harvest swordfish. The experiments proved extremely successful. By the summer of 1963, 90 percent or better of those engaged in the swordfishery had converted to longlining, and harpooning became a secondary harvesting method, so much so that most of the Cape Island–type longliners did not rig with a top mast and substantially shorted their harpooning pulpit.

Unfortunately, this success was short-lived. The increased pressure on the stocks caused by longlining and the fact that a large number of juvenile fish were being caught was starting to become evident in the decrease in the size and number of fish being landed. At this time there was no formal regulation with respect to quota or the size of fish that could be landed. It was basically a free-for-all that would have dire consequences in the coming seasons. By the 1965 season, the individual catches had decreased in excess of 60 percent when compared to that landed three years previous. But the biggest blow was yet to come.

The major market for Nova Scotia swordfish at this time was the United States. In 1966, the United States Food and Drug Administration (USFDA) concluded that the mercury content found in imported Canadian swordfish exceeded American health and safety standards, and they placed a ban on the import of Canadian swordfish. Their decision was based on tests that revealed that swordfish that came from the northern waters off the Canadian coast had levels above the acceptable limit of 0.5 parts per million (ppm), while those tested from southern waters off the eastern coast of the United States had concentrations within acceptable limits. This came as a devastating shock to the Nova Scotia swordfish fishers. They were absolutely bewildered, given the fact that the swordfish in question were caught in international waters and on the same northern fishing grounds as those fished by their American counterparts. In most cases, American and Canadian vessels were fishing almost side by side, yet the Americans' fish were acceptable while the Canadian-caught fish were not. It also came as a shock to some of the major buyers along the eastern coast of the United States, as they were well aware that the American fishers alone could not satisfy the demands of the market and they were facing a considerable loss of revenue. Despite major objection from the Canadian government and others from within their own country, the USFDA remained firm in its resolve to enforce the ban.

In Nova Scotia those hardest hit were the vessels operating out of ports east of Halifax and along the eastern coast of Cape Breton, and as a result a number of fishers sold their longliners and sought other ventures. Despite the ban, some along the South Shore of Nova Scotia continued to engage in swordfishing, selling their catch to American buyers in international waters. Surprisingly enough, in 1968 close to eight thousand tons of swordfish were sold at sea to offshore buyers. This practice brought considerable objection from the United States, and in 1971, after extensive testing, Canada agreed that the mercury in the fish was of concern to the federal Department of Health and implemented a complete ban on the Canadian swordfish industry. This ban remained in place until 1979 when an international body of scientists demonstrated that although the mercury content was higher than that found in some seafoods, it was of no greater threat to the health of the general public than levels found in tuna, marlin, and other similar species that were legally marketed in both countries. Once the ban had been lifted, a small number of vessels, primarily from the South Shore of Nova Scotia, re-engaged in the industry, but the harvesting of swordfish did not again become a predominant fishery until the early twenty-first century. The ban was instrumental in the sell-off of a number of the Cape Island–type longliners.

In the mid- to late 1940s Canada became aware of a relatively new market for species that were never harvested in the coastal waters off Nova Scotia. This market was for species such as yellowtail, grey sole, and American plaice. There was a very lucrative market for these species in Europe, and by the mid-1950s a widely expanding market started to develop in North America. The federal Department of Fisheries and the Nova Scotia Department of Trade, Industry, and Commerce believed that this market had considerable potential and should be investigated. These species are normally found on flat, sandy, and small-pebbled portions of the fishing banks and near-to-shore fishing grounds. They are bottom dwellers and as such are difficult to harvest using the conventional hook-and-line methods of traditional longlining. The most effective harvesting method is employing a drag system similar to the otter trawl used on the larger deep-sea trawlers, but instead of using a door system to keep the net open during the drag, this system was similar to a conventional seine system. This method of fishing—known in Europe as the Danish seine, so-named because it was developed in Denmark—was executed quite successfully using medium-sized fishing vessels similar in length to the Cape Island–type longliner. In 1948 the Fisheries Research Board of Canada commenced a series of trials of this method of fishing using their research vessel the MV *J. J. Cowie*, built in 1943 by the Industrial Shipping Company in Mahone Bay, NS. The trials had three main objectives:

- to examine the feasibility of this method of fishing in the near-to-coast waters off the Maritime provinces;
- to identify areas where this fishery could be sustainably exploited; and
- to support their perceived need for a multipurpose, single design vessel for the shore fishery such as the FV *Shirley Adeline*.

The initial trials concluded that the net designed for the capture of flatfish was very effective in the area of Chedabucto Bay but not in the Gulf of St. Lawrence where the bottom was fairly rough for this type of operation. In Denmark the vessels rigged for this type of fishing were in the forty-to-sixty-foot class. The question that was brought forward was, could this type of fishing be conducted not only by dedicated seiners, but also by medium-sized fishing craft?

In 1951 a commercial trial was conducted in the Chedabucto Bay area using a forty-foot fishing craft. The trial confirmed that Danish seining could be adapted to the smaller shore vessels, but the recommendation was that the ideal platform would be a vessel of between forty-five and sixty feet. It also demonstrated that the smaller craft were in fact more effective with the Danish seine than the mid-sized draggers. The only limitations were the need for smooth bottom and that this fishing method could not be carried out on the rough bottom normally worked by the larger conventional draggers. The catches during the trial were very encouraging, to the point that if other areas along the Nova Scotia coast could be found, a commercial fishery could be quite viable. Further areas were investigated and it was confirmed that a sustainable fishery could be supported.

In 1956 a number of fishers along the Eastern Shore of Nova Scotia, in particular the fishing community of Port Bickerton, actively pursued the feasibility of rigging out a Cape Island–type longliner for Danish seining. The results were very successful, and by 1958 a large portion of the coastal longliner fleet had converted to this fishery. During the trials conducted by the Fisheries Research Board, it was found that the coastal waters off Cape Breton stretching from an area northeast of Sydney, around the northern tip of Cape North, to the waters off the fishing village of Chéticamp were ideal for this fishery. As the fishery grew, the ports of Chéticamp, Glace Bay, Dingwall, and Ingonish became the main centres for this enterprise. Longliners from a number of ports along the eastern and southern shores of Nova Scotia and eastern Cape Breton joined those whose home port was Chéticamp and fished these waters very successfully from early spring until late fall. Until the swordfish ban was put in place, a number of the larger longliners would break away from Danish seining in early June until late August to engage in swordfishing and return in the September time frame. Other areas were found off the eastern shore of Nova Scotia, but these were farther offshore than the fishing areas off Cape Breton. A number of the vessels from the eastern and southern shores of Nova Scotia would continue to fish in these waters, as they were closer to their home ports.

The Danish seine fishery came along at a very opportune time in the life of the shore fishery and the longliner. This new market offset the decline in the groundfish catches from traditional hook-and-line methods and demonstrated the adaptability of the Cape Island–type longliner. This fishery also demonstrated the versatility of the forty-five to forty-eight-foot Cape Island–type longliner that had been gaining popularity. If there was any one drawback this fishery presented, it was that in order to take advantage of the more "fertile" fishing grounds, a good number of the fishers had to operate for extended periods away from their home port. The nature of the species being harvested presented a major problem for the host of small fish plants along the Nova Scotia coast. Most of the plants were built to handle and process traditional groundfish and lobsters and did not have the capability to process sole and other species of flatfish; nor did they have the financial capital to convert. Therefore, longliners from small ports like Whitehead, Marie Joseph, and others had to land their catches either locally in Chéticamp or at ports farther away, such as Canso, Port Bickerton, Halifax, and others that had the processing capability and a collective market. This had a very negative impact on a number of the small processing plants.

By 1964, the major fish plants that were capable of processing the catch were experiencing marketing issues, and a number had a surplus of unsold fish in their storage freezers. It was becoming increasingly difficult for some to sell their catch. In some instances, longliners would return home only to find that the processor could not handle their catch. In such cases it was less costly to just dump their catch than try to travel to other ports in hopes of finding a buyer. As successful as the longliner had been to this fishery, by 1965 the number of longliners engaged in Danish seining started to slowly decline.[93] By the late 1960s, the stocks of grey sole and other bottom-dwelling species, especially in and around northern Cape Breton, also started to decline, and the size and price of the catch compared to the costs associated with the distance to home ports made this fishery unsustainable for some. Other grounds were explored closer to the home ports, but the profitability was less than ideal. The loss of the Atlantic swordfish industry and the steady decline in the groundfish and flatfish stocks once again dealt a major blow to the economic viability of the shore fishery. With the loss of the swordfishery, the longliner fleet suddenly lost one of the fundamental fisheries that was instrumental in its original design and employment.

Despite these setbacks, the Cape Island–type longliner as a class of fishing vessel continued to operate effectively and, in many instances, outperformed the other types of longliners of similar size.[94] By 1973, most of the longliners that had been designed on the dragger or schooner-type hulls had converted to fishing with otter trawl or had entered the scallop fishery. What is surprising is that despite the problems that were prevalent in the shore fishery of the time, longliner construction continued. A small number were replacement vessels for those lost through an accident, but the majority were new construction under the federal subsidy program. Between 1968 and 1972, the construction of the Cape Island–type longliner dropped off significantly, with only nine vessels being launched. There were a number of reasons for this drop in construction that were not related to the decline in stocks and poor market conditions. As was explained earlier, during the time that the yards were constructing longliners and draggers, there was also a high demand for medium-sized seiners for the herring fishery. However, by 1984 the total number of new longliners built increased to fifty-four, 27 percent of the total vessels constructed under the Federal Fishing Vessel Construction Assistance Program.

Throughout the hardships and uncertainties, the longliner soldiered on and established itself as perhaps one of the most versatile vessels to ever engage in the Nova Scotia shore fishery. By 1972 the price for groundfish started to increase somewhat, and a number of fishers returned to longlining, but profit margins were continually getting smaller. A number of the longliners continued to engage in Danish seining, but unlike in the 1960s, catches were getting smaller and a number chose to tie up for the winter months. Some owners chose to rig for gill and driftnet operations while others chose to rig for small-vessel dragging. Some modified their vessels for the scallop industry, while others found their vessels were suitable for employment in emerging fisheries such as snow crab, sea urchin, and sea cucumber.

By the mid-1970s there were ten major fish processing companies in Nova Scotia that combined had become the predominant force in the Maritime shore and offshore fishery. Foremost among them was H. B. Nickerson and Sons Ltd., which bought out National Sea Products Ltd. and alone had the capacity to buy 25 to 50 percent of the Nova Scotia catch. This corporate giant, along with the other nine, continued to expand, building larger and more technologically advanced deep-sea trawlers with the end result being a significant increase in overall fish landing, but the individual landings were gradually getting smaller. A number of the medium-sized plants followed suit, and in their mix of fishing craft, the Cape Island–type longliner still played a prominent role. The scale of the investments made by National Sea Products and others placed them in full control of the Nova Scotia groundfish industry and had a severe impact on the ability of the fleet of individually owned longliners and other smaller craft to show a profit. What is unfortunate is that these major operations had the full support of the federal and provincial governments. Again, very little consideration was given to the shore fishery. The situation had turned 180 degrees, as once again the shore fishery appeared to be in decline. However, the dynamic of the fishery had changed significantly since the dark days of the 1930s, and there was no question it could be sustained albeit with fewer fishers than in the mid-1960s.

In the mid-1970s it became apparent that fish stocks were rapidly declining and that if the industry was to survive, conservation measures had to be implemented. Between 1970 and 1976, Canada worked with the International Commission for the Northwest Atlantic Fishery. In 1972, gear regulations were set, including allowable trawl mesh size, and a quota system was put in place that designated the total allowable catch of specified species in different established geographical zones among the fishing banks off the Canadian coast. The quotas were calculated annually and apportioned among the member countries of the commission.

The 1976 declaration of a two-hundred-mile limit brought some hope and relief to the Nova Scotia fishery. The fishing banks that were subjected to major overfishing by foreign and domestic draggers were now under Canadian control, and access to the banks was governed by the federal Department of Fisheries and was closely monitored. With a number of the major fishing banks now under Canadian control, the major processors aggressively expanded their operations, driving for even greater catches. The rapid expansion of the larger fish processing facilities, the associated increase in dragger activity, and a poor international market caused a surplus of processed fish. By 1979, a number of the smaller fish plants were forced to either cut back their operations or close altogether. Those small plants supported by Cape Island–type longliners that had converted to Danish seining did not have the capability or capacity to process grey sole or other flatfish species, and the fishers were forced to sell directly to the larger producers that were already experiencing overcapacity. This severely reduced output capacity of the small and medium-sized plants and brought into question the overall economic viability of such operations.

By 1976, the United Maritime Fishermen, which had been one of the ten major dealers in the Maritimes, was itself experiencing difficulty. The fishers that once supported the UMF now saw no real advantage in co-operative ownership, and a number of the plants were being bought by the larger operations like National Sea Products. In 1989 the UMF was unable to resolve a number of management issues and went into receivership.

Despite these often detrimental changes in the industry, the long-liner remained a vital component that kept the shore fishery from total collapse. In the early 1980s, the Canadian government realized that if the fish stocks ever were to recover, there needed to be a cap placed on the number of vessels engaged in the fishery, along with a more stringent quota system that would apply to all fishing grounds adjacent to the Nova Scotia coast and under Canadian control. In addition, to protect the inshore fishery from being overfished by the larger vessels, there was a requirement to restrict certain sized vessels from fishing within twelve miles of the coast. To this end the federal Department of Fisheries commenced the implementation of a systematic licensing system that by 1985 included every type of fishing method and species, including lobster. The fishers that were already engaged in the fishery were "grandfathered in," which allowed them to be licensed for the species they had traditionally harvested and the method used. For those that owned and operated longliners, the owner could be granted a licence for longlining, swordfishing, seining, etc., and the licences were applicable to a specific fishing district or the districts in which they had conducted fishing activity before the implementation of the licence system. However, there would be no further licences issued, which meant that to enter the fishery the incumbent had to purchase a licence from an existing owner. In addition, strict quotas were established by species and fishing area. This quota would be reviewed annually and a catch limit would be assigned to each individual fisher. Once this system was in place, the only way to enter the industry was to purchase the quota of a fisher leaving the fishery. The only exception was the opening of a closed fishery or the introduction of a new fishery.[95]

At first glance this policy appears to be very restrictive, but from a shore fisher's perspective, it did provide a degree of protection from larger dragger-type vessels and allowed for better competition with the larger producers. The quota system caused a reduction in the overstocked inventory of fish products and caused a significant increase in fish prices.

What happened in 1992 with the moratorium on the cod fishery is a matter of record. What is interesting is that the longliner and its newer relatives are still plying the fishing banks of Nova Scotia as the lifeblood of the Nova Scotia fishery.

CHAPTER 18

THE LOSSES

They that go down to the sea in ships,
that do business in great waters; These see
the works of the Lord, and his wonders
in the deep.

– PSALM 107: 23-24

THOSE OF US WHO HAVE BEEN BROUGHT UP ON THE SEA KNOW THAT Mother Nature can be cruel and unforgiving. Since the time that people first went down to the sea, there have been losses of ships and crews. Some these tragedies occurred in storms, others by catastrophic events that were complicated by the sheer nature and vastness of the sea, while the causes of others remain unknown.

The loss of any vessel is a tragedy, far more so when human lives are also lost. Fishing in the coastal waters and the fishing banks off Nova Scotia is not immune to the dangers that are inherent with the dynamics of the sea itself. When one adds to this the methods used to harvest the bounty of the sea and the human factors relative to that environment, the risks increase substantially. Accidents occur in every industry, and despite the best technology of the day they still occur. Commercial fishing, be it close to shore or on the distant banks, has always been a dangerous occupation. Over the past sixty years, considerable gains have been made in the area of ships' safety, weather forecasting, training, and technology, yet fishing craft of all types continue to be involved in accidents and incidents that involve the loss of the vessel and/or its crew.

When accidents occur and there is a loss of human life, the first response is to assign blame. Comments like "They should not have been out there in that weather" or "That boat had no place being there—it was a scow" circulate around the wharfs, bars, and wherever sea folk gather to vent. It is a natural human reaction to feel this way, especially when the person voicing their opinion knew or, even more tragically, was related to those involved. Any accident—be it on land, on the sea, or in the air—is a result of factors that when presented either singly or collectively can produce catastrophic results. In a number of coastal villages dotting the Nova Scotia coastline, one can find cenotaphs and monuments that list the names of those who perished at sea. Unfortunately, the Cape Island–type longliner is not excluded from risk factors, and during the forty-six years that are the focus of this study, there was both loss of vessels and, tragically, loss of life.

Many Nova Scotians still remember the tragic loss of the three Cape Island–type longliners from the fishing port of Lockeport on March 18, 1961—the FVs *Marjorie Bryl*, *Johnny and Sisters*, and *Muriel Eileen*—along with a total of seventeen crew members. Amongst those lost were parents, children, siblings, and other close relatives. The tragedy left behind sixty-five children and brought considerable hardship to the unfortunate widows and next of kin as they struggled with the loss of the main breadwinner for each of the families.

This loss followed the loss of the longliner FV *Margaret Lou* out of Lunenburg one year before. These losses added credence to the speculative concern about longliners raised by noted naval architect Howard Chapelle a few years before. Once again questions began to surface as to the safety of the class and whether the government should cancel the subsidy for longliners and concentrate on larger draggers. What many had forgotten is that no class of fishing vessel is immune to such tragedies. In his book *The August Gales*, author Gerald Hallowell provides a graphic and detailed account of the staggering losses that occurred in the hurricanes of 1926–1927. In August 1926 a brutal storm caused the loss of two Lunenburg-based schooners, the *Sylvia Mosher* and *Sadie A. Knickle*; there were no survivors. As if this tragedy weren't enough, in August 1927 four more schooners—the *Joyce M. Smith*, *Mahala*, *Clayton W. Walters*, and *Urda R. Corkum*—were lost along with eighty-five men. These losses sent shockwaves throughout every fishing community along the coast of

The Seamen's Memorial, Canso, NS. (AUTHOR'S COLLECTION)

Nova Scotia. Yet there is no recorded evidence to indicate anyone of authority ever questioned the safety of the schooners or whether they should be operating on the Grand Banks. Similarly, on February 5, 1959, the steel side trawler FV *Blue Wave* out of Grand Bank, NL, sank in a storm with the loss of sixteen souls. The social situation here was as devastating to the community of Grand Bank as were the losses at Lockeport, yet there were no questions raised about the safety or design of the dragger.

Following the Lockeport tragedy, officials from the Department of Transport carried out a formal investigation, and the losses were attributed to a number of environmental factors. For the residents, especially the bereaved, the investigation was unsatisfactory; in their minds it failed to provide anything to verify that the design of the longliner was in fact safe for the employment in which it was engaged. But then, there was no evidence that the Cape Island–type longliner wasn't safe.

The tragedy of any loss always remains in the memory of those that survived or lost loved ones at sea, and there is a tendency to seek someone or something to blame. Such has been the case of the Cape Island–type longliner. Time does not take away the memory of the tragedy. The loss of the longliners out of Lockeport is still fresh in the memory of the residents some forty-five years after the event. What is equally tragic is that it is extremely doubtful any one person could have ably and adequately satisfied the concerns of those who raised them, regardless of what science and/or evidence was presented. In fairness, it might be worthwhile to examine the whole class over its lifetime rather than concentrate on the tragic incidents.

A thorough review of the information contained in the official ship's registers and other records held by Transport Canada revealed that of the 205 wooden Cape Island–type (government approved) longliners built between 1949 and 1985, 106 (51 percent) were lost or destroyed through various accidents/incidents while at sea. Sixty, or 29 percent, were lost to sinking. (For the purpose of this analysis, this includes losses through the ingress of water by catastrophic failure of equipment, collision with an object or another vessel at sea, and loss due to environmental issues.) As concerning this figure may appear, my research revealed that only eight of these losses involved loss of life.[96] In looking at the class as a whole, therefore, it would be statistically correct to state that 8.3 percent of those lost by sinking involved loss of human life. Compared to the losses that occurred in the dragger and trawler fleets at sea during the same time period, the number of lives lost in accidents involving the longliner was significantly lower, and the high survival rate provides some indication that the vessel itself was relatively safe and stable. A review of the vessels that were lost shows that the losses were spread over all builders, and there is no indication that a specific design flaw could be attributed to a specific geographic area or builder. In a majority of the incidents, human factors played a major role. The location of the losses was distributed over a vast geographic area, with the vessels involved operating from different ports both in Nova Scotia and Newfoundland. In addition, there was no concrete correlation or trend to indicate the losses occurred more in one season than another.

The second largest loss of vessels was due to fire. A review of the registry information for each vessel under study revealed that fire at sea contributed to the loss of thirty, or 15 percent, of the vessels built. These figures do not include those lost to fire while alongside, which was treated as a separate category. A closer analysis revealed that there was no particular period of time where the losses due to fire were more predominant. Instead, the losses were spread over the life of the class. What is interesting is that the majority of the fires were determined to have occurred in the engine room, and again, human factors were a component. Two of the losses involved an engine room fire despite having a manually operated, fixed fire suppressant system installed in the space. In both cases the fire occurred while the crew were working on deck, and by the time the fire was discovered, there was no chance of entering the wheelhouse to activate the extinguisher system. This was the only area where one could consider there was a shortcoming in system design; however, there was nothing to indicate problems with the actual vessel construction or design.

Here are some other statistics from the research:

- Fourteen vessels, or 6 percent, were lost as a result of grounding; only one of the incidents involved loss of life.
- Nine vessels were lost to damage while alongside. Most of this damage was as a result of storms, but at least three were lost as a result of a fire to a structure/buildings or wharf adjacent to where the vessel was tied up. Two were lost as a result of an onboard fire, one of which was reported as being suspicious in nature and did not involve the owner or the crew.
- Fifty-nine longliners, or 28 percent, are confirmed as having been removed from service and scrapped. It is worthy to note that the majority of the vessels in this category exceeded their predicted eighteen-year service life, albeit some had undergone major refits and had their hulls coated with fibreglass. For a wooden fishing craft of this size, and with due consideration to the environment in which it was employed, this number speaks well of the design and construction of the vessel.
- Seven of the longliners were sold to international interests and were used as coastal cargo vessels more so than fishing. It can only be assumed that these vessels are no longer in service; however, given that they perhaps ended their life outside Canada they cannot be considered as being lost to any accident or incident.

- A review of the official ship's registry information held by Transport Canada revealed that thirty of the Cape Island–type longliners built between 1949 and 1985 are still active as of the time of writing, in 2022. A number of these vessels were built in 1956 through to 1958, and although most have undergone major refits, rebuilds, and inspections, their survival speaks well of the overall design. If one were to combine the number of vessels still active with the total number of vessels that were known or assumed to have reached the end of their service life, the total figure would be eighty-nine or 43 percent of those constructed. At first glance it would appear to be a relatively small number, but one must not lose sight of the fact that some of the vessels that were built in the early 1950s were nearing the end of their projected service life just as others were coming out of the ship-yards, and yet a good number are still afloat and active. Losses are indeed tragic, but overall the Cape Island–type longliner has proven to be very sound.

EPILOGUE

If ye win through an African jungle,
Unmentioned at home in the press,
Heed it not; no man seeth the piston,
But it driveth the ship none the less.

- FROM THE 1919 POEM "THE LAWS OF THE NAVY,"
BY REAR ADMIRAL R. A. HOPWOOD, R.N. (RET.)

IT HAS OFTEN BEEN SAID THAT IF THE NOVA SCOTIA FISHERY IS NOT IN crisis, then it's going to be. Since the mid-1920s the turbulence that seems to surround this industry appears to ebb and flow in ten-year cycles, and the two primary issues seem to remain constant—declining fish stocks and inaction by both levels of government. Regardless of one's opinion, the shore fishery and the communities it supported never returned to the poverty and desperation of the 1920s, and although reduced in scope for a host of reasons, we still have a shore fishery in Nova Scotia and the other Maritime provinces. The recovery of the shore fishery cannot be attributed to one single factor but to a host of initiatives spanning over three decades. From a social perspective, the work done by Coady and Tompkins allowed the shore fishers to recognize their collective potential as it relates to the formation of co-operatives and the UMF. The fishers were able to garner the support of the press, both national and international, which played a key role in forcing government action to make available funding for the purchase of new fishing craft and equipment, including the longliner. Foremost among this class was the Cape Island–type vessel.

When Rear Admiral Hopwood, RN, first published "The Laws of the Navy" in 1896, he couldn't possibly have imagined how one verse of this epic poem would be so applicable to the Cape Island–type longliner almost fifty-five years later and beyond. The golden age of the wooden Cape Island–type longliner lasted a mere thirty-five years. From its very beginnings, it persevered through considerable political controversy at both levels of government; was basically on trial from the day the first of its class hit the water; could be credited with being the major player that saved a host of small communities all along the Nova Scotia coast from economic ruin; and, of course, contributed significantly to the recovery of the shore fishery itself. Yet, like the verse from "The Laws of the Navy," very little of its dramatic and colourful history has been recorded for the enjoyment and knowledge of future generations. To the public at large, it was just another fishing boat. But to the fishers that knew these vessels, they were much more—they were part of their life.

The last known wooden Cape Island–type longliner that fulfilled all the requirements laid down by the federal government in the 1947 order-in-council that defined the longliner class, the requirements of the Federal Fishing Vessel Construction Assistance Program for the longliner class of fishing vessel, and the provincial standard for that class was the FV *Mr. Co-op* launched from the Cheticamp Boatbuilders yard in late 1984. This vessel was actively engaged in both the Nova Scotia and Newfoundland fisheries until its demise in January 2000. In March 1985 the Federal Vessel Construction Assistance Program was cancelled. Since 1984 there have been a multitude of vessels that were known as "longliners" launched from a host of Nova Scotia yards, and the majority qualified for some form of subsidy that was subject to the changes brought on by a host of amendments introduced particularly after 1965 and programs for smaller vessels introduced following the cancellation of the Federal Fishing Vessel Construction Assistance Program. However, none of these vessels met the true specifications of the Cape Island–type longliner as defined by the federal legislation of 1947 for the "longliner class" of fishing vessel, nor could they be called a government approved longliner.

By the end of 1985, which ends the period covered by this book, there were a multitude of changes taking place both in the fishery and in vessel construction. By the beginning of 1986 new construction

The FV *Passage Pride*, with a home port of Eastern Passage, NS, provides an excellent example of the second generation of Cape Island-type longliners. This vessel was built in 1988 and is constructed of moulded reinforced plastic. It has a registered length of 41.3 feet, a width of 21 feet, and displaces 59.31 tons. Despite its shorter length, its displacement is equal to or greater than most of the original, larger Cape Island-type longliners, but the lines are unmistakable. (AUTHOR'S COLLECTION)

materials such as fibreglass, carbon fibre, and epoxy resins were starting to replace traditional wood materials, and the day of the wooden fishing vessel, like the sailing ships of old, was starting to fade into memory. With the introduction of new materials, the vessels themselves started to change. They were shorter, wider, and deeper, yet the lines of the original Cape Island-type longliner were still quite visible.

Here's one question that still haunts the history of the shore fishery and this noble class of fishing craft: Was the Cape Island-type longliner successful? In other words, did it achieve what it was designed and brought into service for? There are varying opinions, and over time opinions have a way of clouding facts. The Cape Island–type longliner came into service at a time when there was a critical need to save and expand the shore fishery. Starting in the late 1940s, the traditional near-to-shore fishing grounds off the coast of Nova Scotia were showing the effects of the steam trawlers and seiners, but unlike the dark days of the late 1920s, the shore fishers were gaining their independence and collectively becoming competitive in the marketplace. The depletion of stocks on these grounds required the fishers to go farther out to sea, and from a fisher's perspective there was a need for an affordable mid-sized fishing craft that would enable this to happen. Part of the solution came with the development of the Cape Island–type longliner.

Most of the surviving fishers who were involved with the government approved longliner as either owners or crew are now well up in their years. It is their firm belief that this class of vessel could have been in service much earlier had it not been for the federal and provincial political wrangling that dominated the fisheries landscape through the mid-1940s. Some believe that had the provincial government listened to the shore fishers and looked at the Cape Island design rather than trying to develop a common vessel platform for the small draggers, their fishery would have expanded quicker and perhaps there would have been more built during the early '50s through to the mid-1960s. While this might be true, it must be acknowledged that the provincial government had a very difficult task and numerous political obstacles to overcome before the expansion of the shore fishery could get underway in earnest.

In the mid-1970s, which would have been the height of longliner activity but for the loss of the swordfish industry, there were approximately 253 fish buyers and processing firms registered and operating in Nova Scotia. Of the 253, at least 190 (or 75 percent) of them were small or medium-sized local enterprises. When large firms ran their own fleets of offshore draggers and groundfish trawlers, the smaller firms depended on the independent fishers and their vessels. It is estimated that about 35 percent of the supporting fleet were vessels capable of coastal or near coastal longlining, and 40 percent or more of that number were of the Cape Island type. With respect to the groundfishery, 25 percent of the ports in Nova Scotia's coastal fishing villages were supported in some form or another by one or more of the Cape Island–type longliners, and where there was a longliner, there was a small to medium-sized processing facility. Although the vessel's design was approved primarily for the harvesting of ground-fish, between 1947 and 1985 the wooden Cape Island–type longliner was involved in almost every aspect of the fishing industry except for those activities that could only be carried out by a large steel deep-sea trawler. It is not surprising that in 2022, a number of the original Cape Island–type longliners are still active in some aspect of the Nova Scotia fishing industry, while the large steel trawlers are no longer in service and the majority have been sold and/or scrapped. In this regard, the Cape Island–type longliner was very successful, and the fact that it is still visible in the shore fishery throughout the Maritimes and on the West Coast demonstrates the quality of its design and versatility.

In 1977 the extension department of St. Francis Xavier University, Antigonish, NS, tasked Ms. Lawless Saunders to research and prepare a report for the fishers of Nova Scotia that could be used as a reference document to further their collective voice in the fishing industry and to counter the power that they believed to be held by the giant fish companies. This report, entitled "A Rough Voyage," contains some very interesting facts with respect to the role that the longliner had played in the Nova Scotia fishery and what role it could play in the shore fishery of the future. In her presentation Ms. Saunders referred to a report by the Canadian Council on Rural Development that compared the investments made in trawlers to that made for longliners. The figures presented were based on the costs known in 1976. The analysis stated that the average trawler represented an investment of approximately $4 million and was staffed by a crew of about fifteen. Statistics show that the large steel trawlers of the period could land about six million pounds of fish annually.

By comparison, the Cape Island–type longliner of the period represented an investment of $250,000 and was operated on average with a crew of four, including the captain/owner. On average this vessel could land about one million pounds of groundfish per year. Ms. Saunders hypothesized that sixteen longliners, representing an investment equivalent to one of the trawlers, could land up to 2.6 times the amount of fish as a trawler and provide employment for sixty-four people at sea. It was also estimated the trawler created three jobs on land for each job at sea, but this could not be extrapolated to the small vessel fishery. However, in using these figures and the ratio of one job on land for each job at sea, a longliner could theoretically generate 1.4 times the employment on land as would a trawler. Ms. Saunders's report also emphasized that the longliner employing hook-and-line or other similar methods would not have the same impact upon the fish stocks as the trawler and there would be no damage to the grounds and fish habitat. On the other hand, the trawler fishery can ensure continuity of supply, considered a necessity for maintaining regional competitive strength. And the longliner, regardless of the numbers, could not conduct a fishery for a period of time during the winter months.

Unfortunately, changes to the fishery—the limit placed on the number of fishing licences and the implementation of quotas by the federal government from the mid-1970s until the complete closure of the cod fishery in 1992—would not allow any exploration or detailed study of notion. It would have been very interesting to see if the idea of using longliners versus trawlers would have allowed the shore fishery to be sustained even with the quota system. It would be equally interesting to see how many of the smaller fish processing plants, both co-operative and privately owned, would have remained open.

The cancellation of the subsidy in 1986 had considerable impact on the construction of any medium to large-sized fishing vessels, including the Cape Island–type longliner. A small number were built after the subsidy cancellation, but they began to take the name of their proposed employment, that is, dragger, seiner, etc. With the cancellation of the subsidy the classification system used by the federal Department of Fisheries for the qualification of the subsidy ceased to exist, as did the term "government approved" longliner. Changes have been occurring in the shipbuilding industry itself since 1986. Most of the yards that traditionally had constructed the wooden Cape Island–type longliner were either closing or being sold to other builders. Wood was rapidly being replaced by fibreglass and other synthetic composites as the prime construction materials. This building technology allowed builders and architects to introduce changes to the design of the traditional longliner platform. The longliners of today are averaging forty-five to fifty feet in length with a breadth in excess of twenty-five feet. Most are capable of engaging in the lobster and other fisheries and are equipped with state-of-the-art navigation and communication technology. Yet as one wanders around the wharves of the various fishing communities, the lines of the original

Cape Island–type longliners are visible in these new marvels. What is equally interesting is that the design, albeit modified as a result of technology, has replaced the traditional dragger and schooner-type hull forms and has become the most prominent type of fishing vessel on the Atlantic seaboard.

What began as a dream in our ancestors' eyes not only became a reality, but also embedded itself in the very fabric of the history and heritage of Nova Scotia. We owe a debt of gratitude to those brave men and women who had a dream—a dream that their children could flourish from the bounty of the sea—and in realizing that dream created a legacy that deserves to be remembered. We must not forget what these vessels did during their lifetime for the economies of a host of small and medium-sized fishing villages along the Nova Scotia coastline, which in turn affected the economy of the province as a whole.

Although almost forgotten to history, the Cape Island–type longliner played an important role in their lives and successes. This sturdy fishing craft lives on in the designs that are the very fabric of Nova Scotia's shore fishery today. Nova Scotians must ensure this proud heritage is maintained and that future generations understand and respect the ocean that sustains them.

ACKNOWLEDGEMENTS

THIS BOOK WOULD NOT HAVE BEEN POSSIBLE WITHOUT THE SUPPORT of the following (listed in alphabetical order and located in Nova Scotia, unless otherwise indicated): A. F. Thériault & Son, Ltd., Meteghan River; A. L. LeBlanc Ltd., Wedgeport; Acadia University, Wolfville; Acadian Cultural Center, Pubnico; Age of Sail Museum, Port Greville; American Bureau of Shipping, Spring, Texas; Bedford Institute of Oceanography, Dartmouth; Canadian Coast Guard Station, Port Bickerton; Canadian Coast Guard Station, Sambro; Caravan Fisheries Ltd., Ketch Harbour; Cheticamp Boatbuilders, Chéticamp; Coady Institute, St. Francis Xavier University, Antigonish; Dalhousie University, Halifax; Fisheries and Oceans Canada Library, Dartmouth; Fisheries Museum of the Atlantic, Lunenburg; H. Hopkins Ltd., Port Morien; Library and Archives Canada, Halifax and Ottawa; Library of Parliament, Ottawa; Mahone Bay Museum; Maritime Museum of the Atlantic, Halifax; Nova Scotia Archives, Halifax; Nova Scotia Fisheries and Aquaculture Loan Board, Truro; Museum of Civilization, Hull, Quebec; St. Francis Xavier University, Antigonish; Shelburne County Museum; Snyder's Shipyard Ltd., Dayspring; Transport Canada, Board of Steamship Inspection, Dartmouth; Transport Canada, Marine Safety Branch, Ottawa and Dartmouth; Transport Canada Office of Ship Registration, Dartmouth; University of Toronto.

The following Nova Scotia fishers were essential to my research (all the names marked with an asterisk are people who sadly are no longer with us): Captain Vernon Conrad*, Lunenburg; Captain Kevin Fiset*, Grand Étang; Captain Ariel Gray*, Sambro; Captain Keith Horton, Port Bickerton; Captain Claude O'Hara,* Louisbourg; Captain Tom MacDonald, Port Morien; Captain Moise Porier*, Grand Étang; Captain George Ross and Captain Percy Ross*, Stoney Island, Cape Sable Island; and Captain George Toughs, Ketch Harbour.

The following Nova Scotia builders contributed immensely to the content of this book: Bill Cox, Shelburne; Wayne Croft, Dayspring; Roland DesChamp, Shelburne; Cecil Heisler, Indian Point; Philip Snyder*, Dayspring; G. Thériault, Meteghan; Robert Zinc, Bridgewater. And the following Nova Scotians also contributed: Albert Deveau, Chéticamp; Ralph Getson, Lunenburg; Clifford, Glen, and William (Bill) Hopkins, all of Port Morien and all, sadly, deceased; Edward Marcoux, Dartmouth; and John* and Harry Robar, Dayspring.

And finally, my deep thanks to my lovely wife, Theresa, who stood by me as we visited almost every fishing village in Nova Scotia and roamed wharves over a period of five years, spent countless hours in museums and university libraries, and visited the ship registry and other government offices located in Dartmouth and Halifax. For her steadfast support, encouragement, and counsel, I will be forever grateful!

APPENDIX A

AN INITIAL VENTURE IN LOBSTERS, 1929

AT FATHER JIMMY'S URGING, A FEW FISHERS OF WHITEHEAD CAME together in 1929, leading eventually to the formation of the Whitehead Co-operative Society. Billy Tom Feltmate was the first secretary of the group, and in 1937 he spoke at a Co-op Conference about how it all began eight years before. His remarks are recorded in Peggy Feltmate's book, *From Whitehead Harbour, Guysborough County, Nova Scotia: Its Stories, Histories, and Families.* They are reprinted here with Peggy's permission.

We were getting seven cents a pound for market lobster and three cents for small ones. Now we had an idea that the lobsters must be selling at a good price in Boston. But how could we ship lobsters? We poor fishermen didn't even know how to ship a crate: we didn't even know to whom to send it, and we were so used to being skinned that we were afraid to try even one crate.

But after talking the matter over a great many times, we decided to try a crate among the four of us. We said that if 140 pounds of large lobsters make a crate, at seven cents a pound it was $9.80 among the four—and if we lost it, why that was our hard luck! You see, God Almighty knows everything, and when He made the fisherman, He knew that there was a certain class of people out to skin him. So, He took precautions and ordered it so that when one skin was pulled off him, another grew on. It's a good thing He did or 90 percent of the fishermen today would be walking around skinless!

All right, we got a buyer's name in Boston out of the "Fishing Gazette," and we shipped our crate to him. Every evening we met, the four shareholders of the crate of lobsters, and talked the matter over. Some said if we could get $15 it would be much better to ship to Boston. Another said, "I don't ever expect to hear tell of them." That poor fellow had been skinned twice in one year.

Finally one morning, after returning from hauling my traps, the wife met me at the gate and says, "There is a great big envelope come this morning with a picture of a lobster on it." I had never got a cheque for lobsters before, and I didn't know which end of the envelope to open. I didn't know how it happened, never will know, but I struck the right end. There was a cheque for $32.00!

I don't think I'll ever forget the feeling that came over me. I thought there must be some mistake, and one of the chaps suggested that I'd better send the cheque back. As soon as news got around that we received $32 for one crate of lobsters, there was trouble. The packers held a meeting and told us they wouldn't buy the small lobsters if we shipped the large ones. We got together and figured up what we had gained. And… so we organized in what is called a Fisherman's Co-op. Three months after organizing, we brought the price of twine down from 70 cents to 37 cents; laths from $7.50 to $3.50 a thousand; gasoline from 40 cents to 26.5 cents. That's what organization did for us.

APPENDIX B

LOAN BOARD AGREEMENT, SELECTED PAGES

𝕿𝖍𝖎𝖘 𝕬𝖌𝖗𝖊𝖊𝖒𝖊𝖓𝖙 made in triplicate the   14ᵗʰ

day of   October   A.D. 19 59

BETWEEN:

John McLean & Sons Ltd of   Mahone Bay
in the County of Lunenburg
Province of Nova Scotia, hereinafter referred to as "The
Builder,"

OF THE FIRST PART

—and—

FISHERMEN'S LOAN BOARD OF NOVA SCOTIA
a body corporate, having its head office at Halifax in the County
of Halifax, Province of Nova Scotia, hereinafter called "The
Board"

OF THE SECOND PART

— and —

██████████████   of   ██████████████
██████████████

Province of Nova Scotia, hereinafter called "The Fisherman".

OF THE THIRD PART

WHEREAS the Builder has agreed to build and equip a vessel for the Board, and the Fisherman has agreed to purchase the said vessel from the Board when completed, all as hereinafter set forth.

████ AND WHEREAS the Fisherman has made a deposit with the Board of the sum of ████████ against the purchase price of the said vessel.

NOW THIS AGREEMENT WITNESSETH that the Builder and the Board, in consideration of the covenants, agreements and conditions herein contained, mutually covenant, promise and agree to and with each other in manner following, namely:

1. Subject as hereinafter provided, the Builder shall build and equip for the Board at the yard of the Builder at   Mahone Bay   aforesaid, a Fishing Vessel to bear the temporary name   of the following principal dimensions: —

Length Overall   50 feet

Beam (moulded)   15 feet

~~Draft(at rest)~~   Depth   8 feet about

approved by the Board of Steamship Inspection
in accordance with plans and specifications ~~which the Board has supplied to the Builder. The said plans and specifications shall remain the property of the Board and the Builder shall return the same to the Board after the vessel is completed. The Builder shall not make or permit anyone else to make copies of said plans or specifications or anon or disclose for it taking or permit anyone else to take copies from the said plans or specifications.~~

2. The vessel shall be built under cover (unless the Board shall otherwise permit) and shall be equipped and completed in all respects in strict accordance with the requirements of this Agreement, and the whole of the workmanship, and materials and machinery and equipment and things supplied by the Builder shall be first class in every respect, and the vessel shall be seaworthy, tight and of good, staunch and substantial construction.

of all obligations of the Fisherman herein contained, the said Mortgage to be in such form, and to contain such terms and provisions as the Board shall require.

11. And without restricting the generality of the foregoing, the Fisherman, in consideration of having received a subsidy under the terms and conditions of Order-in-Council P.C. 2490 dated the 24th day of May, A.D., 1951 and an agreement pursuant thereto between His Majesty the King in the right of Canada and Fishermen's Loan Board, shall

(a) keep the vessel actively engaged in fishing operations, using a power gurdy, during the fishing seasons of each of five years from the date of issue of an inspection certificate by a Steamship Inspector of the Board of Steamship Inspection.

(b) not sell or charter the vessel during the said period of five years without first obtaining the consent in writing of the Minister of Fisheries.

(c) during the said period of five years, keep the vessel and its equipment insured at his own expense in an amount not less than the amount of the subsidy plus the amount of loan, if any, owing to the Board, loss to be made payable to the Fisherman and the Board as their respective interests may appear.

(d) during the said period of five years, be responsible for complying with all the requirements of the Board of Steamship Inspection and will at all times have on board the vessel a valid certificate of that Board.

(e) during the aforementioned period, keep the vessel and equipment in an efficient and seaworthy condition and permit inspection of same at all reasonable times by persons authorized by the Department of Transport or the Fishermen's Loan Board.

(f) if default is made in any of the terms and conditions aforementioned, repay forth-with to the Board the amount of the said subsidy less one-fifth of such amount for each ull year that the vessel has been engaged in fishing.

(g) give to the Board:

(1) a personal bond giving and binding his or their personal and/or real property, as the case may be, as security for the repayment of any sum or sums paid to the Fisherman under the terms of the aforementioned subsidy, or;

(2) a bond from a recognized guarantee company in Canada, securing the sum or sums paid to the Fisherman under the terms of the aforementioned subsidy, or;

(3) a bond from one or more sureties giving and binding his, its or their personal and or real property as security for the repayment of any sum or sums paid to the Fisherman under the terms of the aforementioned subsidy.

12. This Agreement and everything herein contained shall enure to the benefit of the assignees of the Board. This Agreement shall not be assignable by the Fisherman without the written consent of the Board.

13. Time shall be of the essence of this Agreement.

14. The word Fisherman, wherever used herein, shall, unless the context other wise requires, be deemed to include the heirs, executors and administrators of the Fisherman; and if the Fisherman consists of more than one person, all covenants on the part of the Fisherman shall be deemed to be covenants by such persons jointly and severally, for them-

SCHEDULE "B" TO AGREEMENT DATED THE 14th DAY OF
October , 19 59 , BETWEEN   JOHN MCLEAN & SONS LTD.

FISHERMEN'S LOAN BOARD OF NOVA SCOTIA AND ███████

CERTAIN WORK TO BE EXECUTED AND MATERIALS TO BE
FURNISHED BY THE BUILDER PURSUANT TO THE FORE-
GOING AGREEMENT.

Price shown in Section 9 of the Agreement shall be payment
in full for the vessel, ready for sea, except bedding, messing and fishing
gear and shall include supplying and installing the following:

1 Model HBW-600 (HW-6-M) Cummins Marine Diesel rated 126
continuous B. H. P. at 1800 (175 maximum horsepower) complete with 3:1
hydraulic reverse reduction gear, fuel float tank, 2 fuel filters, 2 lube
oil filters, sump pump, compression release, Penn Safety Alarm, Wheelhouse
Instrument Panel with gauges and tachometer, fresh water corrosion resistor,
heat exchanger cooling and Variable speed governor, f.o.b. Mahone Bay.

1500 watt generator

27 F. P., F. P. T. O. 1:1 ratio

1 set H. D. Batteries

Bronze Tail shaft 2⅜"

Intermediate Steel Shaft

1 - 3-blade Bronze propeller size as recommended by the engine
people.

Installation complete with bearings, stuffing boxes, etc., to
comply with the requirements of C. S. I.

Keel to be of Oak, Yellow Birch or Maple size 8" x 8".

Planking of Spruce 1¾"

Garboards of Oak, Yellow Birch or Maple

Shear Strake of Oak

Oak ice sheathing around water line about 36" in width on the
average.

1 Lunenburg Foundry 2-hole Oil Burning Range No. 12 installed
with pump and gravity tank.

1 mechanical gurdy installed.

Rudder to be of iron, wood filled, to the requirements of the
C. S. I.

Bilge pumping arrangement to be the requirements of the C. S. I.

1 Galvanized Iron Fresh Water Tank to be placed under forecastle
floor, with capacity as space permits.

SCHEDULE "B"

2 Fuel tanks total capacity not to exceed in any case 700
gallons and not less than 600 gallons, as space permits.

1-32 volt ½ mile ray Search light.

1 electric horn

Equipment as outlined in paragraph 36 of the Fishermen's
Loan Board requirements covering legal equipment.

Wire hand rails on forecastle deck

1-13 foot bottom dory

1 mast to be about 7½" at deck placed in after part of boat as
is usual.

Riding sail of size to suit captain

Shoe to be of hardwood 1½" x 8"

Hog Piece to be of Hardwood 3½" x 10"

Gurdy house to be penned off and have lead scuppers.

Running lights in accordance with requirements of C. S. I.

1 stainless steel sink installed in forecastle with ¾"
clock pump to water tank

1 galvanized anchor to comply with C. S. I. requirements. Owner
to supply a second anchor if required.

Forecastle to be laid out to suit the captain or owner.

Steamship Inspection fees cost of trial runs except fuel and
lube oil.

Linoleum on floors of Pilot house and forecastle.

Finishing of pilot house and forecastle to the satisfaction
of the captain.

Fishhold to be divided into pens and have interchangable boards.

Depth recorder)
Radio telephone)   to suit owner but not to exceed $2,000.00
Loran sets   )   total cost.

APPENDIX C

CHRONOLOGICAL LIST OF LONGLINERS BY NAME, REGISTRATION NUMBER, BUILDER, AND DATE OF REGISTRATION

NOTE: THE BASELINE DATA HAS BEEN TAKEN DIRECTLY FROM THE OFFI-
cial vessel register for each individual vessel. In an effort to establish
the sequence of launch, the registration date was used as the baseline. The reasoning for using this date was that in a number of cases
the actual date of the vessel's launch could not be established, and
normally the vessel was formally registered with the Department of
Transport just prior to or shortly after launch but prior to sea trials.

NAME	REGISTRATION #	BUILDER	REGISTRATION DATE
JANET LOUISE	179253	Clarence R. Heisler, Youngs Island, NS	January 1, 1948
ZOOMER	191226	Clarence R. Heisler, Youngs Island, NS	September 1, 1949
ZILCH	191238	Clarence R. Heisler, Youngs Island, NS	January 23, 1950
DAVID PAULINE	192129	Warren Robar, Upper LaHave, NS	June 7, 1950
CLARENCE AND WALTER	170919	Warren Robar, Upper LaHave, NS	January 15, 1951
JUDY AND LINDA	190748	Kenneth MacAlpine & Son, Shelburne, NS	March 29, 1951
PAT & JUDY	193863	Harley S. Cox & Sons Ltd., Shelburne, NS	July 31, 1951
SANDRA AND LINDA	193869	Clarence R. Heisler, Youngs Island, NS	October 25, 1951
MANATEE	193842	Smith & Rhuland Ltd., Lunenburg, NS	November 23, 1951

LADY OF FATIMA	194106	Smith & Rhuland Ltd., Lunenburg, NS	November 26, 1951
EKOLITE	194430	Smith & Rhuland Ltd., Lunenburg, NS	January 14, 1952
BETTY HARRIS	194110	Smith & Rhuland Ltd., Lunenburg, NS	January 30, 1952
MURIEL EILEEN	194482	Harley S. Cox & Sons Ltd., Shelburne, NS	February 22, 1952
LADY LYNN II	194610	John C. McLean & Sons Ltd., Mahone Bay, NS	April 22, 1952
SYLVIA ANN	193850	Clarence R. Heisler, Youngs Island, NS	May 30, 1952
ATLANTIC ROVER	194466	Kenneth MacAlpine & Son Ltd., Shelburne, NS	June 19, 1952
GERTRUDE & RONALD	194465	Harley S. Cox & Sons Ltd., Shelburne, NS	July 14, 1952
GOLDEN NUGGET	194465	Harley S. Cox & Sons Ltd., Shelburne, NS	July 18, 1952
JUDITH & DWIGHT	195018	John McLean & Sons Ltd., Mahone Bay, NS	November 18, 1952
ADA & BILL II	194526	Oscar W. d'Entremont, Lower West Pubnico, NS	December 11, 1952
CHARLOTTE F	195497	Oscar W. d'Entremont, Lower West Pubnico, NS	December 15, 1952
CHARLENE A	176632	Harley S. Cox & Sons Ltd., Shelburne, NS	February 2, 1953
FOAM	194470	Harley S. Cox & Sons Ltd., Shelburne, NS	July 10, 1953
A.B.G.B.	195932	Whitman Kaiser & Sons, Port Bickerton, NS	July 27, 1953
RAY-OLA-K	194471	Harley S. Cox & Sons Ltd., Shelburne, NS.	September 5, 1953
LEYLAND	195944	Irvin Kaizer, Port Bickerton, NS	April 26, 1954
BONNIE GALE	194478	Harley S. Cox & Sons Ltd., Shelburne, NS.	November 3, 1954

LINDA DIANNE	198232	John C. McLean & Sons Ltd., Mahone Bay, NS	November 16, 1954
ERNEST AND RONALD	198296	Oscar W. d'Entremont, Lower West Pubnico, NS	November 24, 1954
WENDY EILEEN	348646	Deschamp & Jackson Boatbuilders Ltd., Shelburne, NS	December 17, 1954
BEVERLEY DOREEN	194709	Clarence R. Heisler, Youngs Island, NS	March 17, 1955
ELIZABETH COLLEEN	198356	Smith & Rhuland Ltd., Lunenburg, NS	April 21, 1955
CAROLYN ANN II	198424	Harley S. Cox & Sons Ltd., Shelburne, NS	April 27, 1955
BETTY & DONNA	198426	Harley S. Cox & Sons Ltd., Shelburne, NS	July 20, 1955
SEADOG	197258	Smith & Rhuland Ltd., Lunenburg, NS	July 20, 1955
RALPHIE J.	198861	Smith & Rhuland Ltd., Lunenburg, NS	August 17, 1955
FAYE AND MOLLY	198429	Harley S. Cox & Sons Ltd., Shelburne, NS	October 17, 1955
ROCKET II	198869	Smith & Rhuland Ltd., Lunenburg, NS	November 1, 1955
HARVEY & SISTERS	188088	Smith & Rhuland Ltd., Lunenburg, NS	November 7, 1955
STELLA MARIS II	188130	Kenneth MacAlpine & Son, Shelburne, NS	November 28, 1955
ELLA & ROBEY	195364	Harley S. Cox & Sons Ltd., Shelburne, NS	December 10, 1955
JOAN OF ARC	198430	Kenneth MacAlpine & Son Ltd., Shelburne, NS	March 3, 1956
ZEPHYR	198431	Harley S. Cox & Sons Ltd., Shelburne NS	April 13, 1956
JANET & MARJORIE	188204	Smith & Rhuland Ltd., Lunenburg, NS	April 19, 1956

SHIRLEY & ARTHUR	188368	Smith & Rhuland Ltd., Lunenburg, NS	May 2, 1956
KENMORE	188187	A. F. Thériault & Son Ltd., Meteghan River, NS	May17, 1956
ISLE ROYALE	188133	John McLean & Sons Ltd., Mahone Bay, NS	May 28, 1956
DOROTHY PEARLE	198432	Harley S. Cox & Sons Ltd., Shelburne, NS	June 11, 1956
MARLENE & SONIA	188132	Smith & Rhuland Ltd., Lunenburg, NS	June 14, 1956
MILDRED L. WHITE	198434	Kenneth MacAlpine & Son Ltd., Shelburne, NS	June 29, 1956
LENNARFISH	189173	John C. McLean & Sons Ltd., Mahone Bay, NS	July 3, 1956
GAYLE YVONNE	310775	A. F. Thériault & Son Ltd., Meteghan River, NS	July 22, 1956
JANE-MARIE	198436	Harley S. Cox & Sons Ltd., Shelburne, NS	July 30, 1956
EDITH JANE	189177	A. F. Thériault & Son Ltd., Meteghan River, NS	September 15, 1956
GEORGE III	189180	John McLean & Sons Ltd., Mahone Bay, NS	October 1, 1956
BETTY & MARION	198437	Kenneth MacAlpine & Son Ltd., Shelburne, NS	October 29, 1956
JOE BOOK	1984381	Harley S. Cox & Sons Ltd., Shelburne, NS	November 6, 1956
MARY SYLVIA P.	189185	Wagstaff & Hatfield Ltd., Port Greville, NS	November 30, 1956
LAURA ELLEN	189313	John McLean & Sons Ltd., Mahone Bay, NS	December 3, 1956
CAROLYN & SHIRLEY	183069	Wagstaff & Hatfield Ltd., Port Greville, NS	January 18, 1957
SEN SEN	198440	Kenneth MacAlpine & Son Ltd., Shelburne, NS	January 21, 1957

R. J. DOUGLAS	189471	Harley S. Cox & Sons Ltd., Shelburne, NS	February 7, 1957
MARGARET LOU	189315	Smith & Rhuland Ltd., Lunenburg, NS	March 25, 1957
MARILYN & STUART	189472	Kenneth MacAlpine & Son Ltd., Shelburne, NS	April 15, 1957
PHYLLIS & RUTH	189314	Smith & Rhuland Ltd., Lunenburg, NS	April 29, 1957
PAYZANT SISTERS	189643	Clarence R. Heisler, Youngs Island, NS	May 9, 1957
I WONDER	189474	Harley S. Cox & Sons Ltd., Shelburne, NS	May 13, 1957
SEABREEZE IV	310185	A. F. Thériault & Son Ltd., Meteghan River, NS	May 13, 1957
JOAN EILEEN	189188	A. F. Thériault & Son Ltd., Meteghan River, NS	May 15, 1957
ORAN II	189744	John McLean & Sons Ltd., Mahone Bay, NS	June 7, 1957
LEONA	189189	Wagstaff & Hatfield Ltd., Port Greville, NS.	June 12, 1957
ST. PETERS	310109	Wagstaff & Hatfield Ltd, Port Greville, NS.	June 25, 1957
WAYNE AND FARLEY	189475	Kenneth MacAlpine & Son Ltd., Shelburne, NS	July 4, 1957
LEASIDE	310087	Harley S. Cox & Sons Ltd., Shelburne, NS	July 16, 1957
ARTHUR ROSS	310102	A. F. Thériault & Son Ltd., Meteghan River, NS	August 2, 1957
ELIZABETH & LEONARD	310101	A. F. Thériault & Son Ltd., Meteghan River, NS	August 14, 1957
ROBERT & BRIAN	189326	Smith & Rhuland Ltd., Lunenburg, NS	September 6, 1957
GORDON SHIRLEY	189647	Wagstaff & Hatfield Ltd., Port Greville, NS	September 11, 1957

ANNE DENISE	310531	Chester Seacraft Ltd., Chester, NS	September 13, 1957
PAT & DAVID	189646	Harley S. Cox & Sons Ltd., Shelburne, NS	September 17, 1957
JEAN ELIZABETH	189106	John McLean & Sons Ltd., Mahone Bay, NS	October 2, 1957
CHETICAMP II	310117	Fraser & Chaisson Ltd., Chéticamp, NS	October 10, 1957
WAVE	189481	Harley S. Cox & Sons Ltd., Shelburne, NS	October 24, 1957
DEBBIE & BRYAN	310536	A. F. Thériault & Son Ltd., Meteghan River, NS	November 8, 1957
GERALD & VERNIE	189478	Kenneth MacAlpine & Son Ltd., Shelburne, NS	November 14, 1957
ALICE & RITA	310537	A. F. Thériault & Son Ltd., Meteghan River, NS	December 5, 1957
DOROTHY AND RUBY	310540	John McLean & Sons Ltd., Mahone Bay, NS	December 30, 1957
MOUREEN ROSE	189483	Harley S. Cox & Sons Ltd., Shelburne, NS	January 9, 1958
KAREN DAWN	310722	Wagstaff & Hatfield Ltd., Port Greville, NS	March 16, 1958
LYNN AND SHEL	189487	Smith & Rhuland Ltd., Lunenburg, NS	June 1, 1958
MARJORIE BYRL	190750	Harley S. Cox & Sons Ltd., Shelburne, NS	June 25, 1958
NEW STAR	310733	Wagstaff & Hatfield Ltd., Port Greville, NS	June 25, 1958
BLAINE & GLEN	310568	Harley S. Cox & Sons Ltd., Shelburne, NS	August 13, 1958
LEAH JOANNE	310564	John C. McLean & Sons Ltd., Mahone Bay, NS	September 11, 1958
FLASH III	310517	A. F. Thériault & Son Ltd., Meteghan River, NS	September 22, 1958

ST. BERNADETTE	310573	Fraser & Chaisson Ltd., Chéticamp, NS	October 27, 1958
ROBIN LYNN	310576	A. F. Thériault & Son Ltd., Meteghan River, NS	December 19, 1958
JOYCE GLORIA	189652	Clarence R. Heisler, Youngs Island, NS	January 29, 1959
CAPE FLO	310025	Herman Doucette, Cape St. Mary's, NS	January 30, 1959
TERRY & GAIL	310785	Smith & Rhuland Ltd., Lunenburg, NS	February 15, 1959
BLUE SWAN	311047	A. F. Thériault & Son Ltd., Meteghan River, NS	March 9, 1959
GAIL AND SHARRY	189496	Harley S. Cox & Sons Ltd., Shelburne, NS	March 20, 1959
GOLDEN NUGGET II	310748	Smith & Rhuland Ltd., Lunenburg, NS	April 10, 1959
JIMMY AND SISTERS	189497	Kenneth MacAlpine & Son Ltd., Shelburne, NS	April 29, 1959
MARGARET AND MARION	311544	John McLean & Sons Ltd., Mahone Bay, NS	April 29, 1959
MACK MARINER III	310085	A. F. Thériault & Son Ltd., Meteghan River, NS	May 4, 1959
DANNY B II	311323	Wagstaff & Hatfield Ltd., Port Greville, NS	June 1, 1959
STE. THERESA	310578	Fraser & Chaisson, Chéticamp, NS	June 4, 1959
NANCY N.	312423	Smith & Rhuland Ltd., Lunenburg, NS	July 3, 1959
E. M. ROY	189654	Clarence R. Heisler, Youngs Island, NS	July 10, 1959
SKY KING	311555	Leonard d'Eon, Lower West Pubnico, NS	July 14, 1959
TERESA AND RUBY	189498	Kenneth MacAlpine & Son Ltd., Shelburne, NS	August 19, 1959

ELAINE JUDITH	312427	Smith & Rhuland Ltd., Lunenburg, NS	September 2, 1959
TOMMY AND JACKIE	311568	John McLean & Sons Ltd., Mahone Bay, NS	October 29, 1959
LUCKY MAE	312973	Ellis Kaiser, Port Bickerton, NS	November 26, 1959
RACER	311578	John McLean & Sons Ltd., Mahone Bay, NS	January 6, 1960
THERESA-HELENE	312445	Fraser & Chaisson, Chéticamp, NS	March 30, 1960,
BETH W.	312699	Ellis and Irving Kaiser, Port Bickerton, NS	May 25, 1960
CAROL & GENE	312721	Smith & Rhuland Ltd., Lunenburg, NS	June 25, 1960
NANCY & LINDA	313645	Smith & Rhuland Ltd., Lunenburg, NS	July 26, 1960
LATOUR	312626	Harley S. Cox & Sons Ltd., Shelburne, NS	June 19, 1961
SANDRA & DIANE	312628	Harley S. Cox & Sons Ltd., Shelburne, NS	August 14, 1961
ST. CHARLES II	312744	Clarence R. Heisler, Youngs Island, NS	November 7, 1961
DOUGIE J.	314322	Fraser & Chaisson Ltd., Chéticamp, NS	February 9, 1962
WAYNE MICHAEL	193845	Smith & Rhuland Ltd., Lunenburg, NS	February 22, 1962
NORTHERN LIGHTS II	314328	Fraser & Chaisson Ltd., Chéticamp, NS	February 27, 1962
ALMA LOUISE	312747	MacLean Shipbuilding Ltd., Mahone Bay, NS	March 7, 1962
FOUR BROTHERS II	319091	Fraser & Chaisson Ltd., Chéticamp, NS	April 16, 1962
MERLIN IV	314525	John E. Clannon, Lower L'Ardoise, NS	June 28, 1962
BONNIE LOU	314981	Clarence R. Heisler, Youngs Island, NS	September 10, 1962

MARGARET & JUNE	195629	Harley S. Cox & Sons Ltd., Shelburne, NS	February 26, 1963
ST. CHARLES III	318915	Clarence R. Heisler, Youngs Island, NS	March 26, 1963
LYNN DENISE	319672	Fraser and Chaisson, Ltd., Chéticamp, NS	August 6, 1963
MISS VALARIE	320852	Fraser and Chaisson Ltd., Chéticamp, NS	February 14, 1963
BONNY LOU II	320640	Fraser and Chaisson Ltd., Chéticamp, NS	March 6, 1964
DEBBIE & MAXINE	320645	Clarence R. Heisler, Youngs Island, NS	April 9, 1964
BARBARA & MARLENE	320864	Berth D. Sampson, L'Ardoise, NS	June 15, 1964
WHITE WHALE	320670	Clarence R. Heisler, Youngs Island, NS	September 24, 1964
CLAUDE & ROY	323056	Fraser & Chaisson Ltd., Chéticamp, NS	September 28, 1964
FLORENCE & NORA	323058	Berth D. Sampson, L'Ardoise, NS	January 8, 1965
SCOTT G.	323103	Fraser & Chaisson Ltd., Chéticamp, NS	May 3, 1965
NELITA AND ROSA	323066	Clarence R. Heisler, Youngs Island, NS	June 11, 1965
FR G. MORNING STAR	314559	Berth D. Sampson, L'Ardoise, NS	August 18, 1965
TRIUMPH	326030	Fraser & Chaisson Ltd., Chéticamp, NS	December 13, 1965
LINDA & MARILYN	325703	Clarence R. Heisler, Youngs Island, NS	February 22, 1966
TOMMIE AND GLADYS	325712	Clarence R. Heisler, Youngs Island, NS	March 28, 1966
TAMMY DARLENE	325714	Fraser & Chaisson Ltd., Chéticamp, NS	August 19, 1966
ROSIE & CALVIN	327430	Clarence R. Heisler, Youngs Island, NS	March 1, 1967

PEGGY ANNE II	326131	Clarence R. Heisler, Youngs Island, NS	July 25, 1967
DECOSTE	327473	Fraser & Chaisson Ltd., Chéticamp, NS	May 6, 1968
KEVIN R.	329816	Ernest L. Rudolph, Marie Joseph, NS	July 9, 1971
B.T.O.	330656	Alfred J. Boudreau, Mavillette, NS	September 23, 1971
PETER BOY	331356	Harley S. Cox & Sons Ltd., Shelburne, NS	October 8, 1971
NATALIE DON II	346432	Alfred J. Boudreau, Mavillette, NS	June 7, 1972
ANNA & ROBEY	346541	Harley S. Cox & Sons Ltd., Shelburne, NS	July 10, 1972
DOUG AND EILEEN	327698	Wagstaff & Hatfield Ltd., Port Greville, NS	August 29, 1972
KIMBERLY & JOYCE	346546	Harley S. Cox & Sons Ltd., Shelburne, NS	October 27, 1972
JEFF & DEAN	327699	Wagstaff & Hatfield Ltd., Port Greville, NS	November 8, 1972
SELENA JEAN	327700	Wagstaff & Hatfield Ltd., Port Greville, NS	March 7, 1973
CHARLES AND SHAWN II	329168	Snyder's Shipyard Limited, Dayspring, NS	March 12, 1973
WHITE CAP III	346548	Harley S. Cox &Sons Ltd., Shelburne, NS	March 26, 1973
ANTHONY AND SUSAN	329169	Snyder's Shipyard Limited, Dayspring, NS	April 17, 1973
PAULA & KRIS	346550	Harley S. Cox & Sons Ltd., Shelburne, NS	May 22, 1973
SHEILA & CATHY IV	346556	Harley S. Cox & Sons Ltd., Shelburne NS	October 19, 1973
KRISTEN AND BJARNI	329175	Snyder's Shipyard Ltd., Dayspring, NS	November 1, 1973
FLASHER	347683	Snyder's Shipyard Ltd., Dayspring, NS	December 21, 1973

SHEILA AND GAIL	346560	Deschamp & Jackson Boatbuilders Ltd., Shelburne, NS	April 14, 1974
SHIRLEY ANN D II	348841	Harley S. Cox & Sons Ltd., Shelburne, NS	April 24, 1974
WINDBREAK	347687	Atlantic Shipbuilding Ltd., Mahone Bay, NS	August 23, 1974
SHEILA PAULINE	347686	Snyder's Shipyard Limited, Dayspring, NS	September 12, 1974
JULIE & BOYS	348648	Harley S. Cox & Sons Ltd., Shelburne, NS	December 4, 1974
TWO DAUGHTERS	347688	Snyder's Shipyard Limited, Dayspring, NS	December 11, 1974
HEIDI & WANDA	348653	Deschamp & Jackson Boatbuilders Ltd., Shelburne, NS	March 20, 1975
PHEASANT	368574	Snyder's Shipyard Limited, Dayspring, NS	September 2, 1975
J. J. SISTERS	370471	Snyder's Shipyard Limited, Dayspring, NS	July 2, 1976
CURTIS & RODNEY	370475	Snyder's Shipyard Limited, Dayspring, NS	August 6, 1976
LITTLE JESSIE	369010	Ernest L. Rudolph, Marie Joseph, NS	August 12, 1976
BAD CAT	370482	Snyder's Shipyard Limited, Dayspring, NS	May 31, 1977
DENIS JIM	370485	Snyder's Shipyard Limited, Dayspring, NS	August 19, 1977
JOHNNY & SISTERS II	383702	Harley S. Cox & Sons Ltd., Shelburne, NS	June 6, 1978
G.T.P.	391594	Harley S. Cox & Sons Ltd., Shelburne, NS	February 13, 1979
CATHERINE G.	392830	Little Dover Boatworks, Little Dover, NS	October 22, 1979
MONICA J.	320852	Snyder's Shipyard Limited, Dayspring, NS	January 10, 1980

LADY DEBORAH	394082	Little Dover Boatbuilders, Little Dover, NS	September 23, 1980
JEFF & TROY	392707	A. L. LeBlanc Ltd., Lower Wedgeport, NS	November 28, 1980
LADY WILIMAE	395951	Harley S. Cox & Sons Ltd., Shelburne, NS	December 30, 1980
MISS CO-OP	801148	Cheticamp Boatbuilders Ltd., Chéticamp, NS.	April 27, 1981
MELODY ROSE III	800304	Harley S. Cox & Sons Ltd., Shelburne, NS	September 1, 1981
KEVIN AND GARY	800446	Cheticamp Boat Builders Ltd., Chéticamp, NS	September 3, 1981
PENNY LANE III	800476	Kaiser Industrial & Marine, Port Bickerton, NS	October 1, 1981
KELLY & DAWN	393563	Clarence R. Heisler and Son, Youngs Island, NS	October 5, 1981
SCOTT COREY	800472	Ernest L. Rudolph, Marie Joseph, NS	October 22, 1981
JONATHAN AND AMY II	801216	A. L. LeBlanc Ltd., Wedgeport, NS	November 10, 1981
SHIRLEY ANN D III	801152	Cheticamp Boat Builders Ltd., Chéticamp, NS	January 6, 1982
NATALIE DON II	393570	Snyder's Shipyard Limited, Dayspring, NS	May 7, 1982
ROSA CARLOS	801346	Little Dover Boatbuilders, Little Dover, NS	May 31, 1982
MARY & NATHAN	801201	Harley S. Cox & Sons Ltd., Shelburne, NS	June 23, 1982
HALF RASPY	802011	Snyder's Shipyard Limited, Dayspring, NS	August 3, 1982
DAVID & JAN	802377	Cheticamp Boatbuilders Ltd., Chéticamp, NS	September 21, 1982
CRYSTAL MARIE	802015	Clarence R. Heisler and Son, Youngs Island, NS	October 26, 1982

WET & WILD	802615	A. L. LeBlanc Ltd., Lower Wedgeport, NS	February 7, 1983
MARILYN G	802018	Snyder's Shipyard Limited, Dayspring, NS	April 18, 1983
DAREN & RANDY	802025	Snyder's Shipyard Limited, Dayspring, NS	September 15, 1983
MR. CO-OP	804903	Chelicamp Boatbuilders Ltd., Chéticamp, NS	January 10, 1984

With the exception of the FVs *Janet Louise*, *Zoomer*, and *Zilch*, all of the above qualified for financial assistance under the Federal Fishing Vessel Construction Assistance Program as defined in 1950. The three aforementioned vessels met all the criteria for the Cape Island–type longliner class of fishing vessel as defined by the federal Department of Transport at the time but were launched prior to the 1950 amendment. From a Department of Transport perspective, these three vessels were the first of their class both in design and structure.

APPENDIX D

FROM THE DIARY OF MASTER SHIPBUILDER
GEORGE EDWARD WAGSTAFF (1887–1978)

THE FOLLOWING TIMELINE IS BY JOHN R. WAGSTAFF WITH ADDITIONAL content extracted from the diaries of George Edward Wagstaff, the master shipbuilder of Port Greville, compiled December 25, 1998. Reproduced with the permission of the Age of Sail Museum, Port Greville, NS.

DECEMBER 31, 1958: Started planking Arthur Beatty's boat "Danny B."

JANUARY 23, 1959: Planking competed, deck all caulked and top sided.

JANUARY 24, 1959: Putting on rail #206 (Arthur Beatty's boat) & paint deck red. Ordered oak.

FEBRUARY 2, 1959: Billy and Ralph putting in floor in accommodations of #206.

FEBRUARY 3, 1959: 6 streaks on at noon, lowering stage this afternoon.

FEBRUARY 5, 1959: Put tank in this morning in #206. Starting to put floor of gurdy house (#206).

FEBRUARY 6, 1959: Finished laying floor in gurdy house & pilot house of #206. 10 streaks of plank on.

FEBRUARY 7, 1959: #206—Put water tanks in this morning.

FEBRUARY 11, 1959: Moved shears back to put in engine for #206.

FEBRUARY 16, 1959: Called about engine for #206—not arrived yet.

FEBRUARY 17, 1959:	Beatty arrived with Clyde Horton who is looking for a longliner.
FEBRUARY 18, 1959:	#206—Truck in Amherst to pick up batteries and have rudder rekeyed.
FEBRUARY 21, 1959:	Harold Puddington down measuring #206 for registration.
FEBRUARY 24, 1959:	Engine arrived for #206.
FEBRUARY 25, 1959:	#206—engine put in boat.
FEBRUARY 26, 1959:	Beatty arrived.
MARCH 2, 1959:	Putting floor in fish hole #206.
MARCH 12, 1959:	Beatty arrived—3 others with him: Mr. Gorman, Mr. Burhoe, and Mr. Campbell this p.m. Kelvin Hughes's man arrived with radar for Beatty boat.
MARCH 14, 1959:	Cheque for 3rd payment on #206 came last night.
APRIL 4, 1959:	Mask all finished for #206.
APRIL 10, 1959:	Launched #206 Beatty boat—"Danny B."
APRIL 14, 1959:	Put swordfishing on #206 this afternoon.

APPENDIX E

NUMBER OF KNOWN GOVERNMENT-APPROVED CAPE ISLAND–TYPE LONGLINERS CONSTRUCTED YEARLY BETWEEN 1950 AND 1985

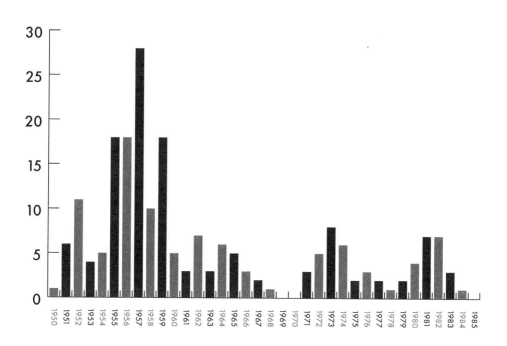

BIBLIOGRAPHY

Admiralty Manual of Navigation, 2 vols. London: Admiralty of the Royal Navy, 1964.

Annual Report and Investigations Summaries 1971. St. Andrews, NB: Fisheries Research Board of Canada, 1971.

Annual Report of the Department of Fisheries 1928–1929. Ottawa: Department of Marine and Fisheries, 1929.

Annual Report of the Department of Fisheries 1933–1934. Ottawa: Department of Marine and Fisheries, 1934.

Barnard, Murray. *Sea, Salt, and Sweat: A Story of Nova Scotia and the Vast Atlantic Fishery*. Halifax: Nova Scotia Department of Fisheries, in cooperation with Four East Publications, 1986.

Bell, James I. *Every Man His Own Shipwright*. London: Chantry Publications Limited, 1950.

Coady, M. M. *Masters of Their Own Destiny: The Story of the Antigonish Movement of Adult Education Through Economic Cooperation*. Halifax: Formac Publishing, 1980. First published 1939.

Choyce, Lesley. *Nova Scotia: Shaped by the Sea, A Living History*. Penguin Group, 1996.

Crosbie, John C. *No Holds Barred: My Life in Politics*. Toronto: McClelland & Stewart Inc., 1997.

Curtis, W. H. *The Elements of Wood Ship Construction*. Almonte, ON: Algrove Publishing, 1919.

Dawson, Robert MacGregor. *Report of the Royal Commission on Provincial Development and Rehabilitation*, vols. 1–2. Nova Scotia: Queen's Printer, 1944.

Duncan, Sir Andrew Rae, W. B. Wallace, and Cyrus MacMillan. *Report of the Royal Commission on Maritime Claims*. Ottawa: Privy Council Office, 1926.

Feltmate, Peggy. "White Head Harbour: Its Stories, History and Families." Self-published manuscript, 2011.

Fisheries and Marine Service. *The Management of Innovation in the Field of Fishing Gear in Canada.* Ottawa: Environment Canada, 1973.

Forbes, E. R. T. "The Origins of the Maritime Rights Movement," *Acadiensis, Journal of the History of the Atlantic Provinces* 5, 1, 1975.

Fraser, M. F. "Note on Changes in the Fishing Industry, Lunenburg, NS, from 1910-1964." St. Andrews, NB: Biological Station No. 995, November 1964.

Gough, Joseph. "History of Commercial Fisheries," *The Canadian Encyclopedia*, August 12, 2013, thecanadianencyclopedia.ca/en/article/history-of-commercial-fisheries.

Government of Canada. "Veteran's Land Act, 1942, A Summary of its Aims Scope and Main Details, Handbook No. 1." Ottawa: Minister of Mines and Resources, 1942.

Hallowell, Gerald. *The August Gales: The Tragic Loss of Fishing Schooners in the North Atlantic 1926 and 1927.* Halifax: Nimbus Publishing, 2013.

Jones, John Henry and Harold Adams Innis. *Report of the Royal Commission, Public Economic Inquiry.* Halifax: Office of the Nova Scotia Provincial Secretary, 1934.

Kemp, Peter. *The Oxford Companion to Ships & the Sea.* London: Oxford University Press, 1976.

Kennedy, William Walker. *Report of the Royal Commission on Price Spreads.* Ottawa: Privy Council Office, 1935.

Lotz, Jim and Michael R. Welton. *Father Jimmy: The Life and Times of Jimmy Tompkins.* Wreck Cove, NS: Breton Books, 1997.

MacDonald, David. "How FX Saved the Maritimes," *Maclean's*, June 1, 1953.

McGuire, Agnes G. "Organizing the Fishermen of the Maritimes," *The Canadian Fisherman*, February 1930.

MacLean, Alexander Kenneth. *Report of the Royal Commission Investigating the Fisheries of the Maritime Provinces and the Magdalen Islands*, 15 vols. Ottawa: Privy Council Office, 1928.

MacNeil, Robert J. *United Maritime Fishermen: Background and Organization.* Antigonish, NS: St. Francis Xavier University, Extension Department, 1945.

Murphy, Liam. *Fishing Cooperatives in Nova Scotia: A Legacy.* Victoria: Seamouse Publishing, 2012.

Nova Scotia Department of Fisheries. *Development of Nova Scotia's Fishing Fleet*. Halifax: Nova Scotia Communications and Information Centre, 1979.

O'Leary, Wayne M. *The Tancook Schooners: An Island and Its Boats*. Montreal: McGill-Queen's University Press, 1994.

Parliament of Canada, House of Commons Debates, 24th Parliament, 1st Session, vol. 1. Ottawa: Library of the Canadian Parliament, 1958.

Parliament of Canada, House of Commons Debates, 24th Parliament, 4th Session, vol. 5. Ottawa: Library of the Canadian Parliament, 1958.

St. Francis Xavier University, Extension Department. *The Antigonish Way*. Antigonish, NS: The Radio League of St. Michael, 1956.

Saunders, Lois. *A Rough Voyage: A Report for the Fishermen of Nova Scotia*. Antigonish, NS: St. Francis Xavier University, Extension Department, 1978.

Serois, Joseph and Newton Wesley Rowell. *Report of the Royal Commission on Dominion Provincial Relations*. Ottawa: Privy Council Office, 1940.

Steward, Robert M. *Boat Building Manual*, 2nd ed. Camden, MN: International Marine Publishing Company, 1980.

Templeman, Wilfred and A. M. Fleming. *Longlining Experiments for Cod off the East Coast of Newfoundland and Southern Labrador, 1950–1955*. Ottawa: Fisheries Research Board of Canada, 1963.

Traung, Jan-Olaf. *Fishing Boats of the World*. Rome: Food and Agriculture Organization of the United Nations, Fisheries Division, 1955.

———. *Fishing Boats of the World 2*. Rome: Food and Agriculture Organization of the United Nations, 1960.

Welton, Michael. "Fraught with Wonderful Possibilities: Father Jimmy Tompkins and the Struggle for a Catholic Progressivism, 1902–1922." University of Toronto, New Approaches to Lifelong Learning Working Paper #57, 2002.

White, Sir Thomas. *Report of the Royal Commission on Financial Arrangements Between the Dominion and the Maritime Provinces*. Ottawa: Privy Council Office, 1935.

ENDNOTES

1   Up until 1976, Canada had jurisdiction over the banks inside three miles of the coastline. In 1976 Canada unilaterally imposed a two hundred mile economic and control zone.

2   Robert J. MacNeil, *United Maritime Fishermen: Background and Organization* (Antigonish, NS: St. Francis Xavier University, Extension Department, 1945).

3   In the early days of the BC fishery, there were ample fish stocks located in the sheltered waters, inlets, and large basins along the coast of the mainland and Vancouver Island. A large percentage of the fishery did not have to venture into the Pacific, whereas a large percentage of the coastal fishery of Nova Scotia was open to the Atlantic.

4   As early as 1867, Nova Scotia fishers adamantly protested the transfer of the responsibility for fishery matters to "Imperial" or Canadian interests. It is indeed ironic that this transfer of responsibilities was done at the request of Sir Charles Tupper, then premier of Nova Scotia. From the outset the fishers thought they were being betrayed.

5   From a brief by Dr. Coady submitted in 1936 to the Honourable J. E. Michaud, minister of fisheries, from the United Maritime Fishermen. Quoted in MacNeil, *United Maritime Fishermen*, 12. A further description of the conditions that existed in the shore fishery of the day can be found in Ken MacLeod, "When cod was king in colonial times, Cheticamp was at the heart of the industry," *Cape Breton Post*, May 4, 2013.

6   Father Tompkins quoted in Jim Lotz and Michael R. Welton, *Father Jimmy: Life and Times of Jimmy Tompkins* (Wreck Cove, NS: Breton Books, 1997), 62.

7   In the mid-1800s, a dispute developed between Canada and the United States over fishing rights, territorial waters, and access to each other's ports for the purpose of getting bait, supplies, etc. In the late 1890s, these issues were finally resolved, with each country able to have limited access to the other's ports. The agreements were called the McCain agreements, and thus the fishing vessels from the US that were granted access to ports like Canso were known as the McCain boats. As a part of the limited vessel access agreements there were certain temporary privileges granted to the visiting boats. These could include buying bait, selling portions of their catch, etc. These privileges were known as the modus vivendi privileges.

8   At this time there were eleven steam trawlers that were operating for Canadian firms out of Nova Scotia ports. The majority of these trawlers were owned by foreign interests and basically leased by Nova Scotia fish processors/dealers. This restriction did not apply to the smaller motorized draggers that were gaining some popularity in the Nova Scotia fishery, particularly along the Bay of Fundy.

9   It is very interesting to note that the annual reports forwarded from the Department of Fisheries paint a very optimistic outlook for the Nova Scotia fisheries. The annual catches for the years 1935–39 show a steady increase in total catches in all groundfish species. The flaw here is that the data is not broken down to show the differences between the inshore and offshore fisheries. In reading these reports one would think that there was no problem with either fishery when in fact the method used to report the year's activity masks the serious problems that existed in the shore fishery.

10 "In effect, the Dominion should be placed on probation for the next few years and if it fails once again in its duty to the fisheries, then the province had better follow the Quebec example and assume virtually complete control." Report of Transmission, *Royal Commission on Provincial Development and Rehabilitation,* 1944, p13.

11 Quoted in Peggy Feltmate, "Father Jimmy, Billy Tom, and the Antigonish Movement." *Early Canadian Life,* September 1980. Peggy Feltmate is William (Billy Tom) Feltmate's granddaughter.

12 Appendix A contains an excerpt from a speech made by Capt. Billy Tom Feltmate to a co-op conference in 1937.

13 By 1964, the small and once impoverished village of Whitehead had expanded their Fishermen's Co-operative to the point where three Cape Island–type longliners were owned and operating out of the port and supporting a local fish plant. Two of the vessels were privately owned, each by a fisher. The third was owned by the Guysborough Co-operative and operated by another local fisher.

14 During this period, there was a lucrative freshwater fishery on the East Coast that centred around species that migrated from the sea to the rivers and lakes. Foremost among the species were gaspereau and shad, both cousins to the herring, that were often used as bait for the inshore lobster industry, and smelts that were sold to local markets.

15 It is a little-known fact that the resentment between the Grand Banks schooner fleet at Lunenburg and shore fishers lasted for decades. Even to this day there are a number of fishers that see the *Bluenose II* as a symbol of the gouging the large dealers did to their fathers and grandfathers and have very little use for it. Fortunately, with the passage of time this issue is becoming less and less relevant.

16 By 1939, newer and better vessels were slowly starting to appear in ports along the coast due in no small measure to the dividends and financial backing provided to the member fishers by the co-operatives.

17 It is worth noting that this average is somewhat swayed by the activity at some of the larger co-operatives.

18 "Loans to Needy Fishermen," *The Canadian Fisherman,* August 1936, 10.

19 "Large Sums Voted to Aid Fish Industry This Year," *The Canadian Fisherman,* September 1936, 78. This board was developed and administered by the Nova Scotia Department of Trade, Industry, and Commerce with no formal act or legislation to provide guidance.

20 By 1936, the positive effects of the co-operative movement and UMF were being felt along the Nova Scotia coast. Fishers were finally getting a better price for their catch, especially lobster, and it was not uncommon for the local fishermen's co-operative to provide some additional financial support for the purchase of better fishing vessels.

21 Honourable Harold Connolly, Minister of Nova Scotia Department of Trade, Industry, and Commerce, Annual Review of Nova Scotia Fisheries, 1949.

22 In 1945, $1,000 would enable a fisher to purchase a new forty-foot, open Cape Island–type fishing vessel ready for sea.

23 Between 1939 and the end of hostilities in 1945, in excess of seventy fishing vessels were appropriated by the federal government and transferred to service with the Royal Canadian Navy. Twenty of these were vessels that were seized from Japanese Canadians after the attack on Pearl Harbour.

24 It is interesting to note that at this time, the restrictions on steam trawlers operating out of Nova Scotia ports was still in effect, and there is no evidence of any legislation that rescinded the original legislation. It would appear that the use of draggers out of Nova Scotia ports was a direct result of the War Measures Act.

25 Many believed that the Federal Fishing Vessel Construction Assistance Program would be repealed upon the cessation of hostilities.

26 A document published by the provincial Department of Fisheries in 1964 listing all fishing vessels forty-five feet and over registered in Nova Scotia ports shows all vessels that did not qualify for the subsidy were designated as "multi purpose" even though some of them were known to have engaged in the groundfishery.

27 The original Bluenose was designed and built without any formal construction regulations or inspections.

28 This directorate was the forerunner of today's Department of Transport Office of Steamship Inspection.

29 Case in point, the American Bureau of Shipping published a set of standards in 1921 that applied only to the United States and its territorial waters. This publication was used as a basis for discussion during the 1943 deliberations described in the next paragraph. With respect to wooden vessels, the building methods used in both the East and West Coast shipyards in Canada played a prominent role in the development of international standards.

30 Lloyd's of London was represented at this meeting and was a major player in the development of the resulting design and construction standards.

31 The foregoing discussion was relative to wooden vessels. There were a number of similar documents produced, each relative to a specific type of ship (e.g., steel tankers, freighters, etc.). By 1951 most of the world's maritime nations had become signatories to these documents; however, how the standards were applied by some nations remains in question to this day.

32 It may surprise some to learn that during the reconstruction of the famous Bluenose II, in the 2011 to 2013 time frame, this 1943 document was used to guide the design and construction.

33 The reasoning for using the tonnage and length was based on the fact that on the East and West Coasts of Canada, there were fishing craft that were thirty-five to forty feet in length yet by their design did not have a registered tonnage at or exceeding fifteen tons.

34 By 1947 Nova Scotia did have a written provincial standard for fishing vessel construction.

35 In some instances there were local builders who served on the boards.

36 The legislation that accompanied the amendment did not refer to the program being open to the Atlantic fishery; however, in July 1942, the federal fisheries minister, the Honourable J. E. Michaud, announced that the subsidy program would be extended to the Atlantic fishery for the construction of draggers.

37 In the Maritime fishery there was an unwritten convention with respect to the definition of trawlers and draggers. Although both used otter trawl, a dragger was normally of wooden construction and below eighty feet in length.

38 At this point, small dragger-type vessels were being examined as a platform for harvesting scallops and herring, primarily from ports along the Bay of Fundy and the Western Shore of Nova Scotia.

39 The legislation defined a group of four or more fishers to include co-operatives and incorporated companies. In the case of the latter, not less than 51 percent of the shares in the company were to be held by not less than four fishing members of the crew of the vessel.

40 This is very interesting as most of the longliners in addition to fishing ground-fish were heavily engaged in the summer swordfishery, yet none were ever challenged.

41 Note that the legislation now applied to a class of vessel, not merely a vessel type.

42 The term "designated fishery" was defined as the fishery or employment for that class of fishing vessel. In the case of the longliner this meant that the vessel could only engage in longlining for groundfish.

43 At this point in Nova Scotia's history, the trawler restrictions were still in effect. The restrictions were formally lifted in late 1944 even though medium-sized trawlers/draggers were being built in Nova Scotia commencing in late 1942.

44 Records indicate that the inspection process took some time to become effective and consistent. It was not until 1947 that a standardized approach was established and qualified inspectors began any substantive inspections.

45 It is indeed ironic that between 1945 and 1951, the Dominion Board Steamboat Inspection Branch had inspected numerous Cape Island–type fishing vessels that were in excess of fifty-five feet in length and met the structural requirements of the longliner as defined in the 1951 subsidy amendment. These inspections found no fault with the approval process used by the Nova Scotia Fishermen's Loan Board.

46 This was not unique to Nova Scotia but was the case for all builders, both on the East and West Coasts.

47 In 1936, longlining was being used with considerable success in the West Coast fishery but was a relatively new fishing method to the Nova Scotia fishery.

48 A. W. H. Needler, *Report of Atlantic Biological Station for 1950* (St. Andrews, NB: Fisheries Board of Canada, 1950).

49 Conversation with Capt. Ariel Fiset, son of the late Capt. Peter Fiset.

50 Between the introduction of the subsidy program in 1947 and the amendment of 1951, only two draggers were built in Nova Scotia yards that satisfied the federal requirements for the subsidy. A number of other vessels were constructed and met the Dominion Board of Steamship Inspection requirements but were not eligible for the subsidy due to the lack of specific provincial design standards.

51 This problem was not unique to Nova Scotia. Documents held by the Fisheries Museum of the Atlantic, Lunenburg, NS, show that the governments of New Brunswick and Prince Edward Island were encountering the same difficulties, and there was considerable dialogue among the three Maritime provinces as to how to develop a common construction standard that could be applied to all vessel types subject to the subsidy.

52 Although the revision was not passed into law until the spring of 1951, the terms of this amendment took effect at the time the revision was tabled in 1950. This is why the *David Pauline*, launched in 1950, qualified for the federal subsidy.

53 As of 1952 and onward, most of the longliners constructed and launched were of single-screw configuration and powered by diesel engines.

54 As explained in chapter 9, in early 1950, the Federal Fishing Vessel Construction Assistance Program was amended. This amendment established new parameters with respect to the length of the vessels eligible for the subsidy and also allowed individual ownership. It was not the tonnage but the length that allowed the *David Pauline* access to the subsidy, and given the "plans" were drawn by a naval architect, it would have met the other requirements. Private builders did not have the latter capability at this time.

55 It is interesting to note that in the late 1940s the federal Department of Fisheries undertook a number of investigations as to the viability of the mechanical trawl hauler on board the MV *J. J. Cowie*. Ironically, the forerunners of the modern trawl gurdy were already seeing service in the fishery, most the design of the fishers themselves.

56 Interview with Capt. Keith Horton, Port Bickerton, NS.

57 Interview with Capt. Keith Horton, Port Bickerton; Capt. George Ross, Stoney Island, Cape Sable Island; and Capt. Claude O'Hara, Louisbourg.

58 Fishers who were engaged in the swordfishery off the coast of Cape Breton in the 1930s using forty to forty-five-foot boats invested in the forty-five to forty-eight-foot boats between 1944 and 1950 then owned and skippered the fifty-five to sixty-foot longliners between 1951 and the late 1960s.

59 In the spring of 1946, the FV *Erdine Mae* was launched at the A. F. Thériault Shipyards, Meteghan River, NS. This vessel had a registered length of 57.7 feet, a breadth 16.8 feet, and a gross tonnage of 34.05. Although the Department of Transport Steamship Inspection Branch classified the vessel as a multipurpose open boat, it was extremely close to meeting the same design criteria as the *David Pauline* that was launched four years later.

60 The small displacement figure for a decked vessel of this size was due to the method used at the time to calculate gross tonnage.

61 I reviewed the registry data of the following vessels: FVs *Carolyn J, Ellenwood II, Mirtland II, Miss Margo II, Oran,* and *Swimm.*

62 The assumption that these other vessels, such as the large snapper boat, were the same as the longliner only bigger, is flawed. From 1944 through to the mid-1950s there were a number of forty-five-foot Cape Island vessels constructed as open boats that were later decked and operated as far out as one hundred nautical miles from the coast. The method of calculating the gross tonnage during the survey would often result in the vessel having a registered tonnage below fifteen when in actual fact it was greater. Information provided by the Canadian Steamship Inspection Branch, Transport Canada, Dartmouth, NS.

63 These types of construction arrangements remain commonplace today. The only difference is that depending upon the size of the vessel, the owner would still be subject to Canadian Board of Steamship Inspection oversight.

64 As a result of previous contracts, the loan board would in most cases already have an approved copy of the builder's drawings in their possession, unless there was a major change in vessel design or configuration.

65 These changes applied equally to the construction or modification of any fishing vessel that qualified for federal subsidy assistance.

66 The initial provincial specification and the federal construction standards did not stipulate the requirement of radio communication equipment or electronic navigational aids. However, most fishers had some communication and navigational equipment installed at the time of construction. As it was part of the contract, it had to be included in the trials.

67 It was not uncommon in some of the smaller boatyards to have the vessels constructed outside. A number of the wooden draggers were allowed to be constructed outside but not the Cape Island–type longliner.

68 By the mid-1960s the ship construction boom had caused a lot of quality native wood species to become scarce, and as a result a number of builders started importing their wood from mills in New Brunswick and eastern Maine. Most builders would only accept wood that was cut between October and April, after which it was allowed to air-dry for an extended period of time before use. In the late 1960s marine grade plywood was approved for use in areas where non-structural sheathing was required.

69 The hog is a fore and aft timber that runs above the keel in a wooden boat.

70 Most Nova Scotia builders used heavier plank than called for in the specification. Some used one-and-three-quarter-inch thickness and used hardwood from the keel to the garboard joint followed by spruce for the remainder of the hull.

71 Industrial-grade marine plywood did not come into use on the longliners until the late 1950s, early 1960s.

72 It is very interesting to note that nowhere in either the original federal construction specifications of 1947 or the provincial design specification is there any reference to the requirement for electronic navigational aids or communication equipment. Nonetheless, all longliners built between 1949 and 1985 were fitted with a host of electronic navigation and communications equipment. Such equipment is now mandatory.

73 In 1977, I underwent navigation training with the Canadian Forces, and the APN 9 was one of the navigation systems still in service aboard the destroyer escort HMCS *MacKenzie*.

74 The cathode ray tube (CRT) was similar to the picture tube found in domestic television sets before the days of digital technology.

75 During the Second World War, the Royal Navy and to a lesser extent the Royal Canadian Navy brought into service a number of coastal patrol/minesweeping craft based on the British steam trawler design. One of their drawbacks was the need for medium-range radar that could be fitted to the ships using the existing power system available on board.

76 It is believed that the FV *Leaside* owned by Capt. Claude O'Hara out of Louisbourg was the first Cape Island–type longliner to have a radar system installed.

77 With the development of solid-state devices and micro electronics, the advances made to the system allowed radar to be installed on open fishing vessels thirty-three feet and longer.

78 During my interview with Capt. Ariel Gray, a well-known and respected fishing captain from Sambro, NS, we discussed the merits of the Decca system. In his view, the accuracy of the Decca Navigator was at times no better than that of the loran, especially in and around the approaches to Halifax.

79 The Loran-C remained the dominant electronic navigational system until the introduction of the Global Positioning System (GPS) in the early 1990s.

80 Master builder Cecil Heisler advised that his late father, Clarence Heisler, had their drawings developed by the naval architect at Lunenburg Foundry and Engineering. They commenced with the approval for the forty-eight-foot longliners and then moved up to the sixty-foot longliners. The process from start to finish took almost two years.

81 In 1960, John MacLean & Sons Ltd., Mahone Bay, NS, alone used 720,000 board feet of lumber in the construction of longliners and draggers.

82 It was not uncommon for shipyards to have more than one vessel under construction at any one time.

83 Appendix E graphs the construction activity from 1950 to 1985.

84 Under the terms of the 1947 amendment to the Federal Fishing Vessel Construction Assistance Program, the minimum allowable length for a long-liner purchased by companies or groups of four fishers was sixty-five feet. The majority of the seven longliners in this subgrouping were constructed under this stipulation.

85 Known by both fishers and builders as the "bastard" dory.

86 Unfortunately, the *Miss Osborne* was destroyed by an explosion in 1951.

87 During one of the many conversations I held with master builder Bill Cox, he confirmed that he was approached by the Province of Newfoundland to establish a yard in Marystown. At the time, the Harley Cox and Sons yard in Shelburne was at full capacity and there was no interest in moving any portion of the operation to Newfoundland.

88 House of Commons debates, 24th Parliament, 4th Session.

89 Ports north of Louisbourg, such as Glace Bay through to Chéticamp, were hampered with drift ice that formed in the Gulf of St. Lawrence between February and the end of March. As a result, a number of the longliners would either cease operations until the ice cleared or move to a location in Nova Scotia south of Louisbourg or operate out of ports on the east coast of Newfoundland. Number of days at sea provided by DFO.

90 In the majority of cases, the owner of the longliner was also the captain.

91 The average wage for those working in industries ashore was $40 to $60 per week.

92 Similar arrangements were declared for the Queen Charlotte Sound, Dixon Entrance, and Hecate Strait on the Pacific coast.

93 In the mid-1960s another form of small-craft dragging, known as Scottish sein-ing, started to appear in Nova Scotia. Its operation was almost identical to the Danish seine except the net had extended wings on each side of the main opening. It achieved some success but had similar problems with fish stocks, as did those that engaged in Danish seine operations. This method never gained the popularity and success of the Danish seine.

94 In the early 1970s, there was a new class of longliner starting to appear on the Nova Scotia coast. This vessel had an overall length of ninety feet. There were also a number of ninety-foot dragger-style vessels being launched as longliners, but most were converted to either otter trawl or scallop operations.

95 The snow crab fishery is a good example of this.

96 The loss of life was as a direct result of the vessel being lost. There were other losses of life through illness or accidents that did not involve the loss of the vessel and as such were not included in this statistical analysis.